Isotopes in Vitreous Materials

Studies in Archaeological Sciences 1

The series Studies in Archaeological Sciences presents state-of-the-art methodological, technical or material science contributions to Archaeological Sciences. The series aims to reconstruct the integrated story of human and material culture through time and testifies to the necessity of inter- and multidisciplinary research in cultural heritage studies.

Editor-in-Chief

Prof. Patrick Degryse, Centre for Archaeological Sciences, K.U.Leuven, Belgium

Editorial Board

Prof. Ian Freestone, Cardiff Department of Archaeology, Cardiff University, United Kingdom
Prof. Carl Knappett, Department of Art, University of Toronto, Canada
Dr. Andrew Shortland, Centre for Archaeological and Forensic Analysis, Cranfield University, United Kingdom
Prof. Manuel Sintubin, Department of Earth & Environmental Sciences, K.U.Leuven, Belgium
Prof. Marc Waelkens, Centre for Archaeological Sciences, K.U.Leuven, Belgium

Isotopes
in Vitreous Materials

Edited by
Patrick Degryse, Julian Henderson
and Greg Hodgins

Leuven University Press

ISBN 978 90 5867 690 0
D / 2009 / 1869 / 1
NUR: 682-971

Lay-out: Friedemann BVBA (Hasselt)
Cover: Jurgen Leemans

Acknowledgements

At the 35th 'International Symposium on Archaeometry' (ISA) in Beijing in May 2005, a number of papers were presented on the use of isotopes in the provenance determination of ancient glass. Also, in many archaeological science journals it was observed that this field was developing. In a discussion with a few scholars, in essence the editors of this volume, it was decided that it would be very timely to put together a state-of-the-art volume on the subject. Taking the idea further, a number of preparatory sessions were introduced at the 36th ISA in Québec and the 17th 'Association Internationale de l'Histoire de Verre' conference in Antwerp in 2006. What lies before you is the result of this process, an edited volume comprising articles by major players active in the field of isotopic research applied to vitreous materials. The subject is topical and is meant to capture the state-of-the-art in this rapidly advancing field. Included are contributions on the origin of and trade in glass and glazes and advances in our knowledge about the part that vitreous materials played in the ancient economy. Besides acknowledging the contributors to all the aforementioned conferences and sessions, warm thanks must go out to Robert Brill, Mark Pollard and Karl-Hans Wedepohl for useful discussions on the compilation of this volume.

Table of Contents

List of Illustrations

List of Tables

Isotopes in vitreous materials, a state-of-the-art and perspectives

Patrick Degryse, Julian Henderson, Greg Hodgins

Introduction

There are many possible research aims behind the scientific examination of archaeological and historical artefacts, but one which has long excited archaeological scientists is the determination of provenance. Such an aim relies on the assumption that there is a scientifically measurable property that will link an artefact to a particular source or production site. Elemental analysis has been used to try and identify where artefacts were produced; it has occasionally been suggested that artefacts could have a compositional 'fingerprint' (Henderson 1989, 31). Indeed some diagnostic chemical compositions of ancient glasses do provide a characterisation which strongly suggests a specific source (Henderson 1988, Brill 1989, Dussubieux *et al*. 2008, Boulogne and Henderson in press). However, although attempts to provide a provenance for glass by elemental analysis continue, a direct relationship between mineral raw materials and the artefacts made from them can be transformed at high temperatures. Archaeological scientists have become increasingly aware that smelting or melting may have had a dramatic effect on the concentrations of many minor and trace elements (Wilson and Pollard 2001, Rehren 2008). Consequently, whereas the study of the provenance of and trade in stone and ceramics is already well advanced, this is not necessarily the case for vitreous materials (Henderson in press). Unlike that for glass and metal, the provenance determination of marbles is often accomplished through chemical and petrographic analysis. Combined thin-section studies and other techniques such as e.g. elemental and stable (carbon and oxygen) isotopic analysis have proven very useful. In 1988 *ASMOSIA* (*Association for the Study of Marble and Other Stones Used in Antiquity*) was founded, to bring together archaeologists, geologists and geochemists to promote a better interpretation of data around stone provenancing. Marble and other stone, especially from the Mediterranean, has received much attention since (e.g. Herz and Waelkens 1988, Waelkens *et al*. 1992, Maniatis 1995, Moens *et al*. 1995, Schvoerer 1999, Herrmann *et al*. 2002, Lazzarini 2002). It is recognized, though, that no one method of analysis alone is sufficient for provenance studies of classical marbles, as overlaps are common when potential sources are compared (Kempe and Harvey 1983). Also in provenance studies of ceramics, since the introduction of petrography and geochemistry in the fifties and sixties (e.g. Peacock 1970), these techniques have been routinely used over the years, especially in the Mediterranean (e.g. Williams 1983). In such studies, the analytical data from the ceramics are compared to the geological characteristics of the potential raw material sources. As with marble, a combination of techniques offers good results. Recent research has also used radiogenic isotopes in the study of marble and ceramics (Brilli *et al.* 2005, Li *et al.* 2006).

However, the precise nature of the raw materials used in glass production and the geographical location where their transformation into prefabricated or completely finished artefacts occurred often remain unclear. In many recent studies, new questions about glass production have been addressed using radiogenic and stable isotopes. The realisation that such transformations as melting have little effect on isotope ratios has provided the basis for the use of this technique. This volume aims to capture the current state of play in this rapidly advancing field when applied to vitreous materials. Included are methodological papers on current, advancing and entirely new techniques for isotope analysis, on the application of new isotope systems to archaeological questions on the origin of and trade in glass and glazes and articles touching on the advances in the knowledge of the ancient economy involving vitreous materials.

Possibilities: relevance of the technique

Different isotopes of an element have the same atomic number but different atomic masses, since they have differing numbers of neutrons. Radioactive decay is the spontaneous disintegration of an unstable radioactive parent isotope to a radiogenic daughter and a nuclear particle. Some isotopes, e.g. those of light elements such as hydrogen, oxygen or nitrogen, have negligible radioactivity and are termed stable. However, a fair number of elements with relatively large atomic masses are radioactive. Such parent and resulting radiogenic daughter isotopes are often used for dating the time of formation of minerals or rocks, but are also very useful in tracing the sources of detrital matter in sedimentary and biogeochemical cycles (Banner 2004). Moreover, variations in many stable isotope ratios reflect different geological origins, due to different formation processes (the original composition of the mineral and subsequent affect of diagenetic or metamorphic processes). The isotopic composition of a raw material is thus largely dependent on the geological age and origin of that material. Conversely, especially the heavy isotopes of e.g.

lead, strontium and neodymium are, due to their relatively high masses at low internal mass differences (Faure 1986, 2001), not fractionated during technical processing. The isotopic composition of the artefact will hence be identical, within analytical errors, to the raw material from which it was derived, while the signatures of different raw materials used, and hence the resulting artefacts, may differ (Brill and Wampler 1965, Gale and Stos-Gale 1982). Isotopes may hence be used in tracing raw materials in craft production… Isotope studies applied to vitreous materials have so far concentrated on four systems: oxygen, lead, strontium and neodymium.

Oxygen and lead isotopes were the first isotopes that were used to investigate ancient vitreous materials. Brill (1970) suggested that oxygen isotopes could be used to provenance the silica fused in ancient glass production. There is a natural range in isotopic values for quartz sands and quartz veins or pebbles depending on their geological origin, e.g. erosion from magmatic rocks, hydrothermal formation, metamorphic conditions. It is stated by Brill (1970) and Brill et al. (1999) that the oxygen isotopic composition of ancient glass will be mainly dependent on the silica source, as it is the predominant component, with minor influences of flux and stabilizer. In experiments, variations in melting time and temperature had limited effects on the oxygen isotope composition of the resulting glass. In a recent study by Leslie et al. (2006), oxygen isotopes were used to indicate that 1st to 4th century AD glass may not have originated from the known region of primary glass production in Syro-Palestine and Egypt, because the signatures of some early Roman glass and 4th-8th century primary glass from the Levant did not match.

Also lead isotopes in a range of archaeological vitreous materials were studied by Brill and co-workers (1974, 1979, 1991) and Barnes and co-workers (1986). In these studies, it was possible to distinguish between vitreous materials in different regions; however, some overlap was noted between 'fields of origin', as was/is not unusual with lead isotopes. In further studies, Brill (1988) observed the lead isotopic compostition of Jalame natron vessel glass (dated to the 4th century AD). Wedepohl et al. (1995) were able to link German medieval wood ash glasses to local ore deposits, while Wedepohl and Baumann (2000) were able to link the lead isotopic composition of natron glass (dated second half to late 4th century AD) from Hambach and Gellep (Germany) to German lead ore deposits. Schultheis et al. (2004) investigated raw natron glass from Ephesos (dated to the 6th century AD), and Degryse et al. (2005) analysed natron glass vessels from Sagalassos (dated to the 5th to 7th century AD). Henderson et al. (2005) analysed some natron and plant ash glass (dated 8th to 9th century AD) from Raqqa (Syria). Freestone et al. (2005) investigated 4th to 8th century AD HIMT glass, and Pernicka and co-workers (El-Goresy et al. 1998) and Shortland (2006) studied the origins of

Egyptian glass. Low lead concentrations in glass, in the 10 to 100 ppm range, are likely to originate from heavy or non-quartz mineral constituents in the glass sand (Wedepohl and Baumann 2000), though low levels of lead have also been ascribed to recycling (Henderson 2000) or to impurities in plant ashes used as an alkali source (Barkoudah and Henderson 2006). The discovery of a homogenous lead isotopic composition in a range of glass samples is likely to suggest that a homogenous silica raw material has been used for the manufacture of that glass, and perhaps that it was produced at a single location (Degryse *et al.* 2005). Heterogeneous lead isotopic compositions of a range of glass samples, particularly with high lead contents, is an indication of recycling of the glass (cf. *infra*), while a linear trend in lead isotopic composition may be indicative of the mixing of two or more end members to produce a glass type (e.g. Degryse *et al.* 2005, Freestone *et al.* 2005). Also, a lead compound may be used for the production of a glaze (Wolf *et al.* 2003; Walton 2005). Though lead isotopes may help to trace sand sources used in glass production, clearly caution is needed in interpreting the results (cf. *infra*).

The application of strontium isotopes in ancient glass depends primarily on the assumption that the strontium in the glass is incorporated with the lime-bearing constituents (Wedepohl and Baumann 2000). It has been assumed that the contribution of natron to the strontium balance of glass is negligible (Freestone *et al.* 2003), and minor contributions may be attributed to feldspars or heavy minerals in the silica raw material (Freestone *et al.* 2003, Degryse *et al.* 2006). Where the lime in a natron glass was derived from Holocene beach shell, the strontium isotopic composition of the glass is that of modern sea water (Wedepohl and Baumann 2000, Freestone *et al.* 2003, Huisman *et al.* in press). Where the lime was derived from 'geologically aged' limestone, the Sr signature in the glass is a reflection of seawater at the time this limestone was deposited, possibly modified by diagentic alteration (Freestone *et al.* 2003). Such signatures can be read from the sea water evolution curve, as defined by Burke *et al.* (1982; see also Dungworth *et al.*, this volume). It has been argued that plant ash glasses are likely to have been produced from low-lime sand or (pure) quartz pebbles (e.g. Henderson *et al.* 2005). In this case, strontium in the glass is derived from the plant ash, the isotope ratio thus reflecting the bio-available strontium of the soil in which the plants grew. This, in turn, reflects the geological origin of the soil (Freestone *et al.* 2003). Both the strontium isotopic ratio and strontium concentrations are useful indicators. Aragonite in shell may contain a few thousand ppm Sr (e.g. Brill 1999). However, conversion of aragonite to calcite during diagenesis or chemical precipitation of calcite or limestone will incorporate only a few hunderd ppm of Sr (Freestone *et al.* 2003). Plant ash glasses can have high strontium contents, sometimes of the same order of magnitude as or higher than glasses made from natron and sand with shell (Freestone *et al.* 2003). Strontium concentrations in plant ash glass often show a considerable variation, indicative of the varied or complex source of the strontium and the concentrating effect of ashing the plant (Henderson *et al.* 2005, Barkoudah and Henderson 2006). Henderson *et al.* (in press, this volume) illustrate how important it is to establish variations in Sr isotope signatures in the landscape as part of trying to provide a provenance for plant ash glasses. The effects of recycling on the strontium isotopic composition of glass have been studied by Degryse *et al.* (2006). With mixing lines, a plot of isotope signatures versus concentrations, it was demonstrated that the Sr in a locally, secondarily produced glass was a mixture of the signatures in two imported end members. A first survey of the Sr isotopic composition of glass throughout the ancient world has indicated the promising nature of the technique in classifying glasses according to their origin (Brill and Fullagar 2006).

The introduction of neodymium isotopes in glass studies is very recent. The Nd in glass is likely to have originated from the heavy or non-quartz mineral content of the sand raw material used. For example the strong correlation between the contents in Nd, FeO and TiO_2 for HIMT glass is an indication of this (Freestone *et al.* in press). Nd isotopes are used as an indicator of the provenance of siliciclastic sediments in a range of sedimentary basin types (Banner 2004). Moreover, the effect of recycling on the neodymium composition of a glass batch does not seem to be significant. Also, the Rare Earth Element (REE) pattern of a glass batch is not expected to change significantly with the addition of colorants or opacifiers (Freestone *et al.* 2005). This offers great potential in tracing the origins of primary glass production. Freestone *et al.* (in press) studied the Egyptian origin of HIMT glass, on the basis of an eastern Mediterranean Nd isotopic signature of the glass. Degryse and Schneider (2008) showed that samples of 1st to 3rd century AD glass could not originate from the known region of primary glass production in Syro-Palestine and Egypt, since the Nd signature of the glass corresponded to a western Mediterranean or north-western European signature. Sr-Nd analysis was also performed on plant ash glass from Banias and Tyre (Degryse *et al.* 2009). Both glass types were made from Levantine coastal sand (with a characteristic high ε Nd) but the two productions used the ashes of plants from different locations (showing a different $^{87}Sr/^{86}Sr$ signature).

Contributions in this volume

This volume consists of a range of articles that reflect the current state-of-the art of the use of isotopes in vitreous materials. In a contribution by Freestone and co-authors, the Sr, O, Nd and Pb isotope compositions of glass from workshops in Syro-Palestine and Egypt are reviewed. It is clear that $^{87}Sr/^{86}Sr$ is particularly useful in the determination of the coastal or inland origin of lime in natron glass and high$^{143}Nd/^{144}Nd$ signifies a strong Nile component, which dominates coastal sands in the region. Pb isotopic compositions are close to the averages for both the Mediterranean Sea and the continental crust; they do not show promise as a general discriminant, but may be useful in specific cases. $\delta^{18}O$ in natron glass ranges between about 13.6 and 15.6 over a number of sites of 4th to 9th centuries AD age in the Levant and Egypt, while in European Roman glass of the 1st to 3rd centuries AD it is typically higher; it is unclear whether this is related to a change in the region of production or is a chronological change related to raw material choice, such as beach sand versus fossil or dune sand. HIMT glass appears to originate in the southeastern corner of the Mediterranean. Glass from the Eifel region of Germany, previously considered to have been made locally, is very close in isotopic and elemental compositions to analysed HIMT from Sinai and is likely to have been made in the same production centre. Plant ash glasses from Banias and Tyre, sharing a coastal signature but differing in $^{87}Sr/^{86}Sr$, have similar $\delta^{18}O$, suggesting that Levantine coastal sand may have a distinctive oxygen isotope signature.

In a study by Degryse and others, the combination of the strontium and neodymium systems is explored. In this case study, the primary origin of early Roman to late Roman and Byzantine glass is evaluated. Production centres of raw natron glass, identified in Egypt and Syro-Palestine, were active from the 1st to 3rd and 4th to 8th centuries AD. Outside these two locations, primary glass units remain unknown from excavation or scientific analysis. The ancient author Pliny tells around 70 AD that besides Egyptian and Levantine resources, also raw materials from Italy, and the Gallic and Spanish provinces were used in glass making. Through the use of Sr-Nd isotopes the primary production of glass in the western Mediterranean or north-western Europe could be proven. Eastern Mediterranean primary glass has a Nile dominated Mediterranean Nd signature (higher than -6.0 ε Nd), while glass with a primary production location in the western Mediterranean or north-western Europe has a different Nd signature (lower than -7.0 ε Nd). Additionally, in comparing the major element analysis and isotope study of these glasses, it becomes clear that both techniques are complementary. While Sr-Nd isotopic analysis may give specific information on the (regional) origin of glass mineral raw materials and their primary origin, this is much more difficult using major elements. For example, the major elemental signature of 2nd century AD Levantine I glasses from the western Roman empire in this study is not indicative of their 'primary' geographical origin, since on the basis of Sr-Nd isotopic analysis of the glass it cannot lie in the eastern Mediterranean.

In a paper by Henderson *et al.* the potential of isotopic research on plant ash glasses is explored. The production and distribution of early Islamic glass in the eastern Mediterranean occurred in centralized urban contexts, within the realm of the caliph and his regional governors. Historical and archaeological evidence in this study suggests that raw glass was manufactured in a small number of centres, such as Aleppo, Damascus, al-Raqqa and Tyre. Scientific analysis of raw and artefact glass has provided evidence of the emergence of a plant ash glass technology in the 9th century AD in the Middle East. In this volume a new approach to the provenance of Islamic glass using radiogenic isotopes is demonstrated. By relating the isotopic composition of this glass to the raw materials used to produce it, it becomes possible to provide a geographical fix for the glasses made in the Islamic world, which should eventually lead to the construction of trade networks for it.

In his contribution, Shortland uses lead isotope determinations of a range of Egyptian vitreous materials from the Late Bronze Age to investigate the relationship between these different materials, many of which have lead as a significant component. Kohl eyepaints are used to give an idea of the lead isotope fields for ore bodies where few data exist. This shows that it is possible to provenance yellow glasses and glazes and to give some information about copper, lead and Egyptian blue pigment provenances. The paper reviews earlier data and adds new trace element data to help interpret the provenance of copper blue glasses and glazes. It shows that the reconstructed patterns are very complex, with the possibility of trace levels of lead being introduced by contamination in the kiln or through one of several components added for either practical or ritual reasons.

Impossibilities: limitations of the technique

Lead isotopes have been extensively used in archaeometry to trace the provenance of metals in the Mediterranean Bronze Age (e.g. Gale and Stos-Gale 1982, Yener *et al.* 1991). The effort, but also the disadvantages and controversy, has been extensively reviewed (for a most recent review see Pollard in press). Because of pitfalls in the use of lead isotopes as a technique, very little new or innovative archaeological interpretation has been published since the mid 1990s. To summarize, issues on accuracy and precision of instrumentation, fractionation and human influence on isotope ratios, the relevance of sampling strategies and the definition of ore fields have demanded much effort. However, many of these questions have been resolved or are much better understood, either by extensive testing or by the technological developments in new instrumentation. More important and dangerous to the progress of the use of isotopic techniques, however, as is clear from the past, may be the lack of mutual understanding, by both archaeologists and scientists, of what can and cannot be proven with isotope systems. This is further discussed under "Perspectives" (cf. *infra*).

A short overview has been given of the current state-of-the-art of the limitations of the four major isotopic systems applied so far to vitreous materials (cf. *supra*). It needs to be stressed that undoubtedly much further research is needed. The lead isotope composition of glass with low lead concentrations may be indicative of the sand used to make it. However, high lead concentrations may be caused by the addition of a mineral compound to the glass batch as a (de-)colourant, possibly strongly influencing or obscuring the original signal of the sand raw material. Moreover, the recycling of glass has typical chemical consequences. The incorporation of old coloured and opaque glass into a batch, with each cycle of re-use, will lead to a progressive increase in the concentration of colourant elements such as copper and antimony, but also lead (Brill 1988, Henderson and Holand 1992, Henderson 1995, Jackson 1997, Freestone *et al.* 2002). Recycled glass may have up to a few thousand ppm of lead, which evidently will entirely alter the lead isotopic composition of the glass, and delete its original raw material signature.

In a paper by Dungworth and others in this volume the nature of strontium isotopic variation is explored and critically reviewed. It is argued, also discussing the use of this system in the investigation of organic materials such as human bone, that some provenance studies based on strontium isotopes are not as secure as they might appear. It is also made clear, however, that strontium contents and isotopic ratios can make significant contributions to the understanding of particular types of raw material used in glass manufacture. In this paper, the use of kelp (seaweed) in the production of 17th century AD glass in the UK is reviewed.

The use of oxygen and neodymium isotopes applied to vitreous materials so far seems to be straightforward. When recycling glass or cullet, the signature of the resulting glass will be a mixture of the end members, but there does not seem to be a factor that can alter or significantly influence the original signal of the silica raw materials. However, it needs to be stressed that the use of both the oxygen and neodymium systems is (still) in its early stages, and pitfalls may arise. It remains to be investigated, for instance, whether the flux in plant ash or wood ash glasses, or the preparation procedure of pure quartz pebbles, may have had an influence on the Nd signature of the artefacts produced (though at least chemical analysis of plant ashes seems to contradict this: Henderson *et al.*, this volume). Also, in particular, the signatures of the raw materials for glass production, be it the silica source or flux, need to be investigated better to understand variations between and within the resources, the relation to the signal retained in the artefacts, and hence the potential of new isotopic techniques for distinguishing primary sources. However, a start has been made in establishing variations in strontium isotope signatures in plants used for primary glass making at one specific Middle Eastern site, and this has been compared with variations in strontium isotope sigmatures across the landscape (Henderson *et al.* in press). Such work underlines the importance of establishing such isotope variations in both the raw materials and the glass.

Accessibility: new techniques

It is beyond the scope of this volume to review the chemistry and analytical methodology of isotope analysis. All principles can be found in e.g. Faure (1986), whereas applications to the earth sciences are reviewed in an excellent way by Banner (2004). The technique and method, however, are relatively new in their application to archaeological raw material research. In summary, analytical procedures in isotope analysis utilise well-established chemical methods of acid dissolution and ion-exchange chromatography designed for the separation and purification of microgram quantities of sample. Using laser ablation, non-destructive micro samples can be investigated, a development with great prospects in archaeological research where sampling an object may be problematic. The chemical preparation of the material sampled where necessary is done in a clean lab environment, to avoid and exclude contamination. Measurements are performed on mass spectrometers, usually high resolution or multi-collector inductively coupled mass spectrometers (HR or MC-ICP) or thermally ionized mass spectrometers (TIMS). New techniques allow a broad application area and a high (fast) sample throughput (Halliday *et al.* 1998). Nevertheless, access to analytical facilities for isotope studies in archaeology may be restricted and/or expensive. This is a critical aspect of what could be termed *isotope archaeology* in material studies. A critical mass of laboratories and research groups investigating isotopes in vitreous materials and other artefacts should be established, to allow for a broad base of research, enough input of new materials and new applications of different isotope systems, and in particular new ideas and inter-laboratory comparison. In particular a synergy with the disciplines of chemistry and earth sciences would be effective in this respect, and would facilitate the development of a revived (post-Pb) field in *isotope archaeology*.

Two contributions to this work explore new instrumental techniques for making isotope analysis more accessible. In a contribution by Marzo and others, the glazes on Islamic and Hispano-Moresque ceramics are studied, and it focusses on the determination of lead isotope ratios by Inductively Coupled Plasma Quadrupole Mass Spectrometry (ICP-QMS). In this way, production areas and periods are distinguished. In the Iberian Peninsula, lead-glazed ceramics were already used by the Romans as imported objects, but it was probably after the 8th century AD, when the Muslims arrived, that this type of ceramic was widely manufactured there, continuing into the medieval and post-medieval periods. The analytical methodology proposed by Marzo *et al.* is simpler than that used in classical Pb isotopic studies, but has sufficient precision to attain the aim of the study, thus opening up the technique to many users. The lead isotope results for Iberian medieval and post-medieval ceramic glazes demonstrate the possibility of establishing a manufacturing pattern for different historical sites and in various periods, as well as indicating the potters' use of different lead ingredients for producing tin-opacified glazes or transparent glazes.

In a study by Walton and others a new method for obtaining Pb isotope ratios with laser ablation time-of-flight ICP-MS is described. To realize precise and accurate isotope ratios, an empirical modelling approach is utilized wherein interference effects, peak tailing, and plasma space-charge effects are inferred by partial least squares linear regression analysis. As a test case to evaluate the precision and accuracy of the model, lead isotope ratios are measured for the lead glazes in Roman ceramics. Since TIMS had previously been used to analyse these glazes, this material provides a point of reference for understanding the performance of the time-of-flight instrument (TOF). It is demonstrated that laser ablation TOF achieves decent precision and accuracy for an *in situ* technique (~0.05% relative standard deviation). As part of a pilot study on the origin of lead used to fabricate Roman lead glazed ceramics of the 1st-2nd centuries AD, isotope ratios were also measured for glazes for which TIMS determinations had not been carried out. For all the glazes measured, close matches were found with Spanish ore sources.

Perspectives

Ancient artefacts are not simply raw material for classification by archaeologists or evidence of production, but they can reveal how productions evolve (Gosden 1994). The study of the different stages of (artefact) production from the acquisition of raw material to the final abandonment of the objects is described as a *chaîne opératoire*. The virtually unexplored field of vitreous materials can provide a unique, interesting and challenging perspective. From the current state of knowledge it is unclear whether glass and glaze may be compared to or contrasted with, for instance, lithics and ceramics when it comes to technological decisions.

However, with the specific information provided by the isotopic techniques described here, archaeologists may understand the origin (and in some cases the nature) of the raw materials used in glass and glaze production and can begin to contribute to models of glass trade, to (socio-)economic models involving their manufacture, and may eventually contribute to models of the ancient economy. Ideally, the interpretation should combine isotopic results with evidence from archaeological excavations, historical sources and vessel form. However, the risk that one or the other research community will internalize its discussion is real, and should be avoided. For almost two decades, only three or four international groups focussed on lead isotope studies applied to the origin of early bronzes, lead and silver dominated the debate on some critical issues in Mediterranean archaeology, and in so doing generated scepticism, not so say controversy (Pollard, in press).

A significant question that has concerned archaeological scientists for some time in this very technical isotopic research is how social and economic questions can be translated into a geochemical research programme and how the results thereof can be incorporated into a model of past human behaviour (Pollard in press). A first step towards resolving this issue might be not restricting the geochemical application, using isotope analysis, to a handful of users. Such situation would control the questions being asked and, even more negative, the interpretations of the results (e.g. Pollard, in press). Secondly, when dealing with isotopes in vitreous materials, the questions that can be answered need to be clarified by the archaeological scientist, and at the outset the archaeological question should be clear. These concerns have always formed part of archaeological science research, including the quality of communication between the archaeologists, the scientists and with those whose expertise combines both areas. The observation that 'although moving away from the original concept of provenance, the relatively simple detection of change in the material record ... is valuable information when interpreting that record' made by Pollard (in press) is significant. The variation in lead isotopic signature of ores, for instance, is a well-known problem in provenance studies: such signatures can be 'broad' and isotopic populations of distinct ore districts may considerably overlap (e.g. Gale and Stos-Gale 1982, Yener et al. 1991), making it impossible to uniquely identify specific ore types used. However, if the archaeological question concerning provenance relates not to a statement of origin but of which one of a number of origins can be ruled out this is, by itself, a valuable contribution.

Applied to glass, an analysis programme focusing on variations in the isotopic composition of a specific vessel form, using well dated samples from areas where the vessel is presumed to have been made, may be much more informative than a random sampling of shapes from one site. Also, in the archaeological science approach, it should be clear that artefact deposition is the result of their use by people and that arefacts had a meaning in ancient socities. Studies on glass production and distribution are generally not integrated into wider (socio-)economic studies. Examplary for the Roman economy, McCray and Kingery (1998) expressed the wish to begin studies on the industrial organization, infrastructure requirements and nature of the value of Roman glass. It is well possible that the absence of glass studies within such wider socio-economic reconstructions is mainly due to the absence of basic data, but it is certainly also due to a lack of integration of archaeological and archaeometrical studies. For a true reconstruction of glass economy, including the context in which people produced these artefacts, such integration is imperative. In connection with socio-economic studies, analyses and graphs should not just be 'pretty' but also 'useful'. The use of a new approach like isotopic studies, now rapidly developing, must obviously be placed in its archaeological context. Such approach must include a consideration of context and date; both are crucial aspects in vitreous materials research.

In conclusion, it is clear that isotopes do not (yet) provide the panacea which is often hoped for. They do not provide a 'fingerprint' that will allow all artefacts to be traced to a unique source. However, as we hope is clear from this volume, many new insights into vitreous materials are revealed when the right questions are asked. Moreover, new approaches are developing fast and, when enough critical mass can be achieved, isotopes applied to vitreous materials can undoubtedly provide new information for archaeologists and archaeological scientists for the next few decades.

References

J.L. Banner, 2004, Radiogenic isotopes: systematics and applications to earth surface processes and chemical stratigraphy, Earth Science Reviews, 65, 141-194.

Y. Barkoudah, J. Henderson, 2006, Plant ashes from Syria and the Manufacture of Ancient Glass: Ethnographic and Scientific aspects, Journal of Glass Studies, 48, 297-321.

I.L. Barnes, R.H. Brill, E.C.Deal, 1986, Lead isotopes studies of the finds of the Serçe Limani shipwreck, in: J.S. Olin, J.M. Blackmann (eds.) Proceedings of the 24th International Archaeology Symposium, Smithsonian Institute Press, 1-12.

S. Boulogne, J. Henderson, in press, Indian glass in the Middle East? Medieval and Ottoman glass bangles from Tell Abu Sarbut and Khirbat Faris, Jordan: typology, texts, scientific analysis and sources, Journal of Glass Studies, 2009.

R.H. Brill, 1970, Lead and oxygen isotopes in ancient objects, Philosophical Transactions of the Royal Society, 269, 143-164.

R. H. Brill, 1988, Scientific investigations of the Jalame glass and related finds, in: G.D. Weinberg (ed.) Excavations at Jalame, Site of a glass factory in Late Roman Palestine, Missouri Press, 257-294.

R.H. Brill, 1989, Thoughts on the Glass of Central Asia with Analyses of Some Glasses from Afghanistan, in: Proceedings of the XV International Congress on Glass, The International Commission on Glass, 19-24.

R.H. Brill, J.M. Wampler, 1965, Isotope studies of ancient lead, American Journal of Archaeology, 71, 63-77.

R.H. Brill, P.D. Fullagar, 2006, Strontium isotope analysis of historical glasses and some related materials: a progress report, paper presented at the 17th international conference of the Association Internationale de l'Histoire de Verre, 4-8 September 2006, Antwerp.

R.H. Brill, I.L. Barnes, B. Adams, 1974, Lead isotopes in some Egyptian objects, in: Bishay, A. (ed.) Recent Advances in the science and technology of materials, Plenum Press.

R.H. Brill, K. Yamazaki, I.L. Barnes, K.J.R. Rosman, M. Diaz, 1979, Lead isotopes in some Japanese and Chinese glasses, Ars Orientalis, 11, 87-109.

R.H. Brill, I.L. Barnes, E.C. Jeol, 1991, Lead isotopes studies of early Chinese glasses, in: J.H. Martin (ed.) Proceedings of the Archaeometry of Glass sessions of the 1984 International Symposium on Glass, Beijing, 65-83.

R.H. Brill, T.K. Clayton, C.P. Stapleton, 1999, Oxygen isotope analysis of early glasses, in: R.H. Brill, Chemical analyses of early glasses, Corning Museum of Glass, 303-322.

M. Brilli, G. Cavazzini, B. Turi, 2005, New data of $^{87}Sr/^{86}Sr$ ratio in classical marble: an initial database for marble provenance determination, Journal of Archaeological Science, 32, 1543-1551.

W.H. Burke, R.E. Denison, E.A. Hetherington, R.B. Koepnick, H.F. Nelson, J.B. Otto, 1982, Variation of seawater $^{87}Sr/^{86}Sr$ throughout Phanerozoic time, Geology, 10, 516-519.

P. Degryse, J. Schneider, J. Poblome, Ph. Muchez, U. Haack, M. Waelkens, 2005, Geochemical study of Roman to Byzantine Glass from Sagalassos, Southwest Turkey, Journal of Archaeological Science, 32, 287-299.

P. Degryse, J. Schneider, U. Haack, V. Lauwers, J. Poblome, M. Waelkens, Ph. Muchez, 2006, Evidence for glass 'recycling' using Pb and Sr isotopic ratios and Sr-mixing lines: the case of early Byzantine Sagalassos, Journal of Archaeological Science, 33, 494-501.

P. Degryse, J. Schneider, 2008, Pliny the Elder and Sr-Nd radiogenic isotopes: provenance determination of the mineral raw materials for Roman glass production. Journal of Archaeological Science, 35, 1993-2000.

P. Degryse, I.C. Freestone, J. Schneider, S. Jennings, 2009, Technology and provenance study of Levantine plant ash glass using Sr-Nd isotope analysis, Archaeologische Korrespondenzblaetter, in press.

L. Dussubieux, C.M. Kusimba , V. Gogte, S.B. Kusimba, B. Gratuze, R. Oka, 2008, The trading of ancient glass beads: new analytical data from south Asian and east African soda-alumina glass beads, Archaeometry, 50, 797-821.

G. Faure, 1986, Principles of isotope geology, 2nd ed, John Wiley and Sons.

G. Faure, 2001, Origin of igneous rocks, the isotopic evidence, Springer.

I.C. Freestone, M. Ponting, J. Hughes, 2002, The origins of Byzantine glass from Maroni Petrera, Cyprus, Archaeometry 44, 257-272.

I.C. Freestone, K. A. Leslie, M. Thirlwall, Y. Gorin-Rosen, 2003, Strontium isotopes in the investigation of early glass production: Byzantine and early Islamic glass from the Near East, Archaeometry 45, 19-32.

I.C. Freestone, S. Wolf, M. Thirlwall, 2005, The production of HIMT glass: elemental and isotopic evidence, in: Proceedings of the 16th Congress of the Association Internationale pour l'Histoire du Verre, London, 153-157.

I.C. Freestone, P. Degryse, J. Shepherd, Y. Gorin-Rosen, J. Schneider, in press, Neodymium and Strontium Isotopes Indicate a Near Eastern Origin for Late Roman Glass in London, Journal of Archaeological Science

N.H. Gale, Z. Stos-Gale, 1982, Bronze Age Copper Sources in the Mediterranean: a New Approach, Science, 216, 11-19.

A. El Goresy, F. Tera, B. Schlick-Nolte, E. Pernicka, 1998, Chemistry and Lead Isotopic Compositions of Glass from a Ramesside Workshop at Lisht and Egyptian Lead Ores: A Test for a Genetic Link and for the Source of Glass, in: C.J. Eyre (ed.) Proceedings of the Seventh International Congress of Egyptologists, Orientalia Lovaniensia Analecta 82, 471-481.

C. Gosden, 1994, Social being and time, Blackwell.

A. Halliday, D.C. Lee, J.N Christensen, M. Rehkomper, W. Yi, X. Luo, C.M. Hall, C.J. Ballentine, T. Pettke, C. Stirling, 1998, Applications of multiple collector-ICPMS to cosmochemistry, geochemistry an palaeooceanography, Geochimica Cosmochimica Acta, 62, 919-940.

J. Henderson, 1988, Electron probe microanalysis of mixed-alkali glasses, Archaeometry, 30, 77-91.

J. Henderson, 1989, The scientific analysis of ancient glass and its archaeological interpretation, in: J. Henderson (ed.), Scientific Analysis in Archaeology, Oxford University Committee for Archaeology, 30–62.

J. Henderson, 1995, Aspects of early medieval glass production in Britain, in: Proceedings of the 12th Congress of the International Association of the History of Glass, 26-31.

J. Henderson, 2000, The Science and Archaeology of Materials, Routledge.

J. Henderson, in press, The provenance of archaeological plant ash glasses, in: A.J. Shortland, Th. Rehren, I.C. Freestone (eds) From mines to microscope - Studies in honour of Mike Tite, University College London Press.

J. Henderson, I. Holand, 1992, The glass from Borg, an early medieval chieftain's farm in northern Norway, Medieval Archaeology, 36, 29-58.

J. Henderson, J.A. Evans, H.J. Sloane, M.J. Leng, C. Doherty, 2005, The use of oxygen, strontium and lead isotopes to provenance ancient glasses in the Middle East, Journal of Archaeological Science, 32, 665-673.

J. Henderson, J. Evans, Y.Barkoudah, in press, The roots of provenance: glass, plants and isotopes in the Islamic Middle East, Antiquity.

J. Herrmann, N. Herz, R. Newman (eds.), 2002, ASMOSIA 5, Interdisciplinary studies on ancient stone, Proceedings of the fifth international conference of the Association for the Study of Marble and Other Stones in Antiquity, Archetype Publications.

N. Herz, M. Waelkens (eds.), 1988, Classical marble: geochemistry, technology, trade, NATO ASI Series E, Applied Sciences, 153, Kluwer Academic Publishers.

D.J. Huismans, T. De Groot, S. Pols, B.J.H. Van Os, P. Degryse, in press, Compositional variation in Roman colourless glass objects from the Bocholtz burial (The Netherlands), Archaeometry, 2009.

C. Jackson, 1997, From Roman to early medieval glasses. Many happy returns or a new birth?, in: Proceedings of the 13th Congress of the International Association for the History of Glass, 289-302.

D.R.C. Kempe, A.P. Harvey, 1983, The petrology of archaeological artefacts, Clarendon Press.

L. Lazzarini (ed.), 2002, Interdisciplinary studies on ancient stone – ASMOSIA VI, Proceedings of the sixth international conference of the Association for the Study of Marble and Other Stones in Antiquity, Bottega d'Erasmo Aldo Ausilio Editore.

K.A. Leslie, I.C. Freestone, D. Lowry, M. Thirlwall, 2006, Provenance and technology of near Eastern glass: oxygen isotopes by laser fluorination as a compliment to Sr, Archaeometry, 48, 253-270.

B.P. Li, J.X. Zhao, A. Greig, K.D. Collerson, Y.X. Feng, X.M. Sun, M.S. Guo, Z.X. Zhuo, 2006, Characterisation of Chinese Tang sancai from Gongxian and Yaozhou kilns using ICP-MS trace element and TIMS Sr–Nd isotopic analysis, Journal of Archaeological Science, 33, 56-62.

Y. Maniatis, N. Herz, Y. Basiakos (eds.), 1995, The study of marble and other stones used in Antiquity, Archetype Publications.

P. McCray, D. Kingery, 1998, The prehistory and history of glassmaking technology, American Ceramic Society.

L. Moens, P. De Paepe, M. Waelkens, 1995, A multi-disciplinary contribution to the provenance determination of ancient Greek and Roman marble artefacts, Israel Journal of Chemistry, 35, 167-174.

D.P.S. Peacock, 1970, The scientific analysis of ancient ceramics: A review, World Archaeology, 1, 375–389.

M. Pollard, in press, What a long strange trip it's been : lead isotopes and Archaeology, in: A.J. Shortland, Th. Rehren, I.C. Freestone (eds.) From mines to microscope - Studies in honour of Mike Tite, University College London Press.

Th. Rehren, 2008, A review of factors affecting the composition of early Egyptian glasses and faience: alkali and alkali earth oxides, Journal of Archaeological Science, 35, 1345-1354.

G. Schultheis, T. Prohaska, G. Stingeder, K. Dietrich, D. Jembrih-Simbürger, M. Schreiner, 2004, Characterisation of ancient and art nouveau glass samples by Pb isotopic analyses using laser ablation coupled to a magnetic sector field inductively coupled plasma mass spectrometer, Journal of Analytical Atomic Spectrometry, 19, 838-843.

M. Schvoerer (ed.), 1999, Archéomatéiaux – Marbres et autres roches, Actes de la IVème conférence internationale de l'Association pour l'Étude des Marbres et Autres Roches Utilisés dans le Passé, Centre de Recherche en Physique Appliquée à l'Archéologie, Presses Universitaires de Bordeaux.

A.J. Shortland, 2006, The application of lead isotopes to a wide range of Late Bronze Age Egyptian materials, Archaeometry, 48, 657-671.

M. Waelkens, N. Herz, L. Moens (eds.), 1992, Ancient stones: quarrying, trade and provenance – interdisciplinary studies on stones and stone technology in Europe and the Near East from the prehistoric to the early Christian period, Acta Archaeologica Lovaniensia, Monographiae 4, Leuven University Press.

M. Walton, 2005, A materials chemistry investigation of archaeological lead glazes, Doctoral Thesis, University of Oxford, Linacre College.

K.H. Wedepohl, I. Krueger, G. Hartmann, 1995, Medieval lead glass from north Western Europe, Journal of Glass studies, 37, 65-82.

K.H. Wedepohl, A. Baumann, 2000, The use of marine molluscan shells in the Roman glass and local raw glass production in the Eifel area (Western Germany), Naturwissenschaften, 87, 129-132.

D.F. Williams, 1983, Petrology of ceramics, in: D.R.C.Kempe, A.P.Harvey (eds.) The petrography of archaeological artefacts, Clarendon Press, 301–329.

L. Wilson, A.M. Pollard, 2001, The provenance hypothesis, in: D.R. Brothwell, A.M. Pollard (eds.) Handbook of Archaeological Sciences, Wiley.

S. Wolf, S. Stos, R. Mason, M.S. Tite, 2003, Lead isotope analyses of Islamic pottery glazes from Fustat, Egypt, Archaeometry 45, 405-420.

K.A. Yener, E.V. Sayre, E.C. Joel, H. Özbal, I.L. Barnes, R.H. Brill, 1991, Stable lead isotope studies of Central Taurus ore sources and related artefacts from Eastern Mediterranean Chalcolothic and Bronze Age sites, Journal of Archaeological Science, 18, 541-577.

Isotopic composition of glass from the Levant and the south-eastern Mediterranean Region

Ian C. Freestone, Sophie Wolf, Matthew Thirlwall

Introduction

The Levant and Lower Egypt have a particular significance in our understanding of early glassmaking. Pliny, writing in the first century AD, identified the Levantine coast as the region where glass was discovered and as the source of early Roman glass (Freestone 2008), and the region continued as a centre of the glass industry through to the enamelled Islamic glass of Damascus in the Middle Ages. Furthermore, it is here that much of our evidence for *primary* glassmaking, the manufacture of glass from its raw materials, has been found. Large scale tank furnaces, yielding clear evidence of glass manufacture but unassociated with evidence of vessel fabrication, have been found dating from the 1st century BC to the 3rd century AD in the Wadi Natrun region in Egypt (Nenna *et al.* 2000, 2005), from the 6th to 8th centuries AD in Israel (Gorin Rosen 1995, 2000, Tal *et al.* 2004) and from the 10th to 11th centuries AD at Tyre, Lebanon (Aldsworth *et al.* 2002). This evidence has given rise to the increasingly accepted division of production model, discussed in many publications (e.g. Nenna *et al.* 1997, Nenna 2007, Foy *et al.* 2000, 2003a,b, Freestone *et al.* 2000, 2002a,b). According to this model, the elemental and isotopic compositions of glass artefacts are determined by the factory in which the primary glass was made, rather than the workshop in which the vessel was fabricated (Fig. 1.1). It follows that if we are to understand the origins of glass in the first millennium AD and beyond, an ability to fingerprint primary glass from the Levantine region is crucial. This chapter reviews our understanding of the isotopic composition of Levantine glass, based upon studies of the past few years. Although it is very probable that glass was made in the region in the Late Bronze Age and Early Iron Age, isotopic studies of this material have largely been restricted to the lead isotope compositions of colourants, and these will not be considered in this paper, which will focus on glass from the Roman period through to the Middle Ages.

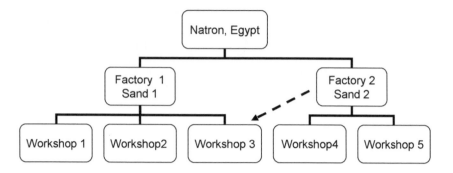

Fig. 1.1
Schematic diagram showing how many workshops may have made glass objects using glass from a single primary glassmaking factory, while a single workshop could make glass using more than one primary composition (from Freestone *et al.* 2008b)

Raw materials

The glasses discussed are all of the soda-lime-silica variety. They fall into two well-known categories. *Natron glass,* a soda-lime-silica type with MgO and K_2O each below about 1.5%, was produced from a mixture of mineral soda and calcite-bearing sand. *Plant ash glass* fused the soda-rich ash produced by burning halophytic plants with lime-poor sand or pulverized quartz to give a glass with around 3% or more K_2O plus MgO. MgO and K_2O contents of the glasses from assemblages discussed below are shown in Fig. 1.2, clearly demonstrating the compositional gap between natron and plant ash glasses. Some early Iron Age glasses, as well as a small number from the period under discussion, do not fall so easily into a specific category, but are not directly pertinent to this paper.

The compositions of sands in the south-eastern Mediterranean are dominated by the sedimentary load of the River Nile. Until the construction of the Aswan Dam, the Nile Delta was advancing into the Mediterranean due to the millions of tons of particulate material carried by the Nile every year. The Nile sedimentary material was carried round the south-eastern Mediterranean coast by tidal currents and longshore drift (e.g. Stanley and Wingerath 1996). The coastal sediments of the eastern Mediterranean are therefore dominated by sediment derived from the Nile. This was demonstrated very clearly by heavy mineral analysis of beach sands which showed the assemblages of the Levantine coast to be dominated by clinopyroxene and amphibole derived from the volcanic rocks of the Ethiopian Highlands (Emery and Neev 1960, Pomerancblum 1966). The volcanic materials are supplied to the Nile by two of its tributaries, the Blue Nile and the Atbara River, whereas the other major tributary, the White Nile, carries negligible volcanics and little pyroxene.

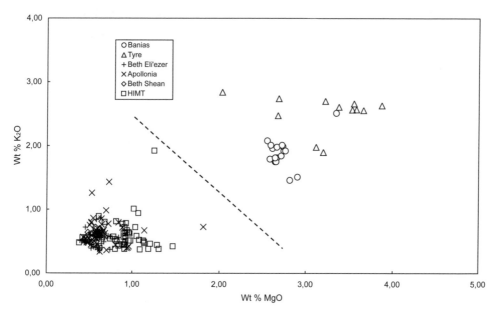

Fig. 1.2
Potash and magnesia contents of glass groups discussed in Freestone *et al.* (this volume), showing the clear division between natron and plant ash types

The coastal sands of the Levant may also contain substantial calcium carbonate, in the form of fragments of shell and *kurkar*, which is a fossil beach sand cemented by calcium carbonate. In an archaeological context, several elemental analyses of sand have been carried out by Brill (1988) and Valotto and Verità (2002) from the Bay of Haifa, the location of good glassmaking sand according to Pliny, demonstrating their suitability for glassmaking. Reworking of sedimentological data (Freestone 2006, 2008) shows that many sands between the Nile and Akko have sufficient calcium carbonate (calcite or aragonite) to produce a glass with 4-10% CaO, but that those of the Bay of Haifa have ideal contents of around 15% $CaCO_3$, yielding a glass with around 8% CaO, and also have a low colour index, so that they are likely to be among the sands with the lowest iron oxide contents in the region.

While pebbles of white vein quartz are widespread; alumina contents of 1 wt. % or more in glasses of the period under consideration suggest that sand was the generally used source of silica. Some of the alumina in Levantine sand is likely to have been present in the form of feldspar; a correlation between potash and alumina observed in the glass from a tank furnace in Apollonia-Arsuf (Fig. 1.3; Tal *et al.* 2004) suggests that potassic feldspar was a source of alumina and potash. K-feldspar is known to accommodate strontium and lead, and its presence in Levantine sand is likely to have influenced isotopic compositions in some cases (cf. *infra*).

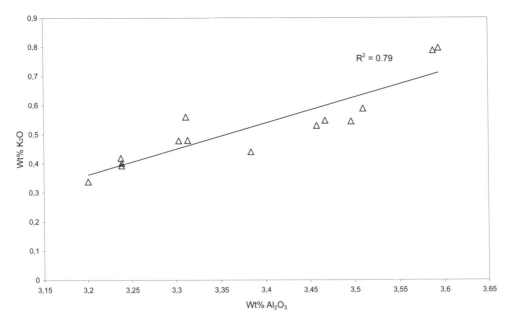

Fig. 1.3
Potash versus alumina contents in glass from a primary glassmaking furnace at Apollonia-Arsuf (Tal *et al.* 2004)

The source of soda in natron glass is generally accepted to have been assemblages of soda-rich carbonates, sulphates and chlorides from evaporitic deposits in Egypt, principally the Wadi Natrun, but also from locations in the Delta such as al-Birnuj (Shortland 2004, Shortland *et al.* 2006). Material dominated by *trona*, a sodium bicarbonate mineral, is likely to have been preferred. Concentrations of Nd, Sr and Pb in these deposits, known collectively as the mineralogically incorrect *natron*, were insignificant relative to those in the sands.

Trace element analysis of plant ashes from Syria have recently been reported by Barkoudah and Henderson (2006). These suggest that while the neodymium contributed to a glass by plant ash is likely to have been of minor significance relative to the sand, plant ash may make a very significant contribution to the strontium concentration, and is likely to dominate the strontium isotopic composition of plant ash glass.

Strontium isotopes

The strontium isotope compositions of Levantine glasses were considered by Freestone *et al*. (2003). Samples from four assemblages of glass were analysed. Two groups of natron glass from Israel, one originating in the primary glassmaking furnaces of Bet Eli'ezer (Hadera; 6th to 8th centuries AD) and one in a secondary workshop at Bet Shean (7th century AD) were compared with a group of natron glasses from Tel el Ashmunein, an 8th to 9th century AD secondary glassworking site in Middle Egypt. A group of plant ash glasses, in the form of primary chunks from a secondary working site at Banias, Israel, were also considered. Results are shown in Fig. 1.4 in terms of total strontium concentration and isotope ratio.

The Sr content of the glasses from the coastal region of Israel is relatively high at 300-500 ppm and $^{87}Sr/^{86}Sr$ approaches 0.7092, the value of Holocene seawater. These data are considered to reflect the derivation of the bulk of the strontium from molluscan shell fragments in the beach sand used to make the glass. The isotopic composition reflects that of the sea water in which the molluscs grew. This interpretation has recently been supported by the analysis of $^{87}Sr/^{86}Sr$ in several sand samples from the Bay of Haifa, near the mouth of the River Belus of antiquity, which approach seawater compositions, dependent upon their shell content (Degryse and Schneider 2008). Unlike the glasses from the tank furnaces, however, which typically have $^{87}Sr/^{86}Sr$ somewhat less than the seawater value of 0.7092 (Fig. 1.4), the sands have somewhat higher values.

The glasses from Ashmunein, which correspond to Group 7 of Foy *et al*. (2003a; also Group 2 of Gratuze and Barrandon 1990), are believed to have been produced in Egypt. These have lower Sr isotope ratios of around 0.7080, consistent with the use of limestone as a source of lime, probably derived from a sand in which particles of limestone were present (although a deliberate addition of limestone cannot be dismissed). The low elemental concentrations of Sr in these glasses (around 150 ppm) support this view, as Sr concentrations in limestone, which have equilibrated with groundwater during diagenesis, are lower than those in marine shell. Finally, the plant ash glasses from Banias have high Sr contents, reflecting the high capacity of plants to take up this element, plus low $^{87}Sr/^{86}Sr$, reflecting the terrestrial origin of the plant ash. Sr contents and Sr-isotopes are clearly potentially of great use in the understanding of glass source materials, as originally proposed by Wedepohl and Baumann (2000).

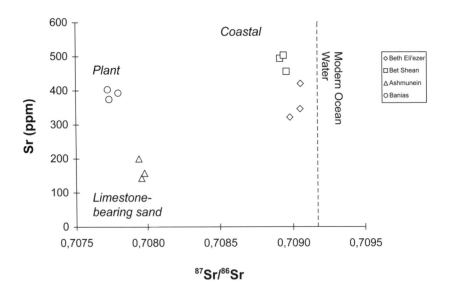

Fig. 1.4
Sr isotopes and total strontium in glasses from various Near Eastern sites; the probable raw materials are indicated (Freestone *et al.* 2003)

Strontium isotopes have also been measured in 8th to 9th century AD glass from Raqqa in Syria by Henderson *et al.* (2005). The natron glasses from Raqqa have isotopic compositions which reflect a marine origin as the source of lime, and the raw glass is likely to have originated in the Mediterranean coastal strip, as anticipated from its composition (Freestone *et al.* 2000). The Raqqa glass is very similar in major element composition to glasses from primary furnaces in the major city of Apollonia-Arsuf and may have originated there (Freestone *et al.* 2008a)

Strontium isotope data for plant ash glasses from 10th to 11th century AD tank furnaces at Tyre in Lebanon have been published by Leslie *et al.* (2006) and also by Degryse *et al.* (in press). Like the Banias glasses, these show a terrestrial Sr signature, but may be distinguished from the Banias production.

Neodymium isotopes

Neodymium isotopic compositions have been reported by Freestone *et al.* (in press) for three raw glass chunks from the tank furnaces at Bet Eli'ezer and a single sample from Apollonia (Arsuf). ε_{Nd} values of between -5.0 and -6.0 are higher than those of most sediments, including those from the Western Mediterranean, reflecting the volcanic component anticipated from the Nile sedimentary load (cf. *supra*). Similar isotopic signatures are found in plant ash glass from Banias, while glass from the tank furnaces at Tyre has ε_{Nd} from -3.2 to -5.6 (Degryse *et al.* in press). Most of these values are significantly lower (more negative) than those for the Nile sediments themselves, which have ε_{Nd} as high as -1.9 (at Rosetta: Tachikawa *et al.* 2004).

Heavy minerals in eastern Mediterranean beach and dune sands have been considered by Emery and Neev (1960). The heavy mineral assemblages of the Nile are dominated by pyroxene and amphibole, and these minerals are also likely to be the major carriers of rare earth elements including neodymium. The pyroxene is mainly derived from Ethiopian volcanic rocks, via the Blue Nile and the Atbara, whereas amphibole is largely derived from metamorphic terrains and is particularly high in the White Nile (Foucault and Stanley 1989). The volcanic signature carried by the sands is therefore likely to be due to the pyroxene. However, as the Nile sediments are moved around the coast of the Mediterranean the heavy minerals are fractionated due to density sorting in the marine sedimentary environment, and they are diluted by mineral assemblages transported to the coast from other regions, such as the Sinai. This is likely to be especially reflected in the pyroxene concentrations in the sands, as the pyroxene component is likely to be derived mainly from the distant volcanic sources and will not be added to in any significant quantity. Fig. 1.5 shows pyroxene concentrations in beach sands around the southeastern Mediterranean, as a proportion of total heavy minerals, from data given by Emery and Neev (1960). It can be seen that there is a marked drop off in pyroxene content away from the Nile, and that the Nile neodymium signature is likely to be diluted accordingly. Hence, it is quite understandable that ε_{Nd} in Levantine glasses can be significantly lower than in the Nile load itself, although it is still high relative to most sands.

The model outlined to explain the lower ε_{Nd} of the Levantine glasses would be convincing, were it not for the analysis of several sand samples from the Bay of Haifa by Degryse and Schneider (2008), yielding higher ε_{Nd} of -4.8 to -1.0, the latter suggesting almost undiluted Nile sediment. Clearly ε_{Nd} does not decrease monotonically around the eastern Mediterranean coast. Furthermore, there are a range of resources for glassmaking, including dune sands and fossil sands such as *kurkar*, which contain less pyroxene than sands from the beach (Emery and Neev 1960), and are likely to have been used at Bet Eli'ezer, where the furnaces are some 10 km away from the sea, and also Apollonia, where the furnaces are on the *kurkar* ridge.

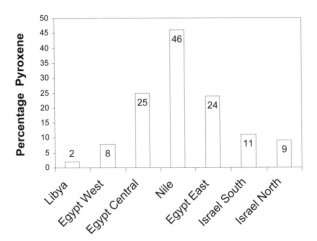

Fig. 1.5
Percentage of pyroxene in heavy mineral assemblages of beach sands from the south-eastern Mediterranean plotted from data compiled by Emery and Neev (1960)

The wide variation in Nd isotopic signatures of glass produced at Tyre is of interest. All of the glasses are chunks, apparently made in the Tyre furnaces, presumably using local sands. Tyre is situated on an isthmus made up in part of a large accumulation of sand (Emery and George 1963, Nir 1996). The variation in ε_{Nd} is likely to reflect the use of sands obtained from different strata. Variation in the isotopic composition of the sands on the Levantine coast is to be expected, reflecting fluctuations in sand supply from the different tributaries of the Nile and from other rivers around the coast, due to climatic fluctuation. Chronological fluctuations in isotopic, elemental and heavy mineral composition are found in cores of Nile Delta sediment, due to climate-controlled variations in the sources of the material (Foucault and Stanley 1989, Krom *et al.* 2002).

The implications of the results are that, on the basis of the small numbers of analyses so far conducted, Levantine sand has a wide range of ε_{Nd}, potentially from -1.0 to -6.0 or lower. Many more data are needed to assess the range of compositions of sands suitable for glassmaking. Whether these may have ε_{Nd} lower than -6.0 is not yet clear, but in order to have diluted the Nile signature of around -1.0 down to -6.0, an older neodymium source would appear to have been contributing to the signature and must have been available in some form.

Oxygen isotopes

The use of oxygen isotopes as a tool in the investigation of early glass was pioneered by Robert Brill, who included a group of glass from a fourth century workshop at Jalame, Israel, in his sample (Brill 1970, 1988, Brill *et al*. 1999) and provided complementary analyses of natron and sand from the Bay of Haifa. In his classic investigation, Brill (1988) showed that the Jalame glass with $\delta^{18}O$ of 14.0 ‰ – 15.5 ‰ was a good match for that expected for a glass made from Egyptian natron and Palestinian coastal sand. Further analyses were carried out by Leslie *et al*. (2006), using laser fluorination of small samples to release the oxygen. They analysed glasses from a number of the sites already discussed: natron glasses from Bet Eli'ezer (Israel), Bet Shean (Israel) and Tel el Ashmunein (Egypt), and plant ash glasses from Banias (Israel) and Tyre (Lebanon). The natron glasses showed similar compositional ranges, with $\delta^{18}O$ ranging from 13.6 ‰ – 15.6 ‰. This, with the Jalame data of Brill (1988), appears to suggest that glasses manufactured from natron and a broadly Egyptian or Nile-derived sand source have an oxygen isotope signature in or around this range. Leslie *et al*. (2006) noted that a limited number of analyses of $\delta^{18}O$ in Roman glass from Europe showed higher values (Fig. 1.6) and that this suggested that some European glass was not made from Palestinian sand, as had been proposed, for example, by Nenna *et al*. (1997) and Picon and Vichy (2003). Further work on the O-isotope signatures of Roman glass by Freestone and Lowry (forthcoming) confirms this pattern.

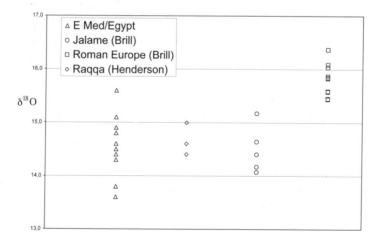

Fig. 1.6
Oxygen isotope data for natron glasses from the eastern Mediterranean and Egypt (4th–9th century AD, Leslie *et al*. 2006), Raqqa (8th–9th century AD, Henderson *et al*. 2005), Jalame (4th century AD, Brill 1988) and Roman Europe of the 1st–3rd centuries (Brill *et al*. 1999)

Recent work on neodymium isotopes of Roman glass by Degryse *et al.* (2008) confirms that the source of Roman glass was not confined to the Levant, but the origin and significance of the differences in $\delta^{18}O$ observed here are not yet fully clear. It may be that the phenomenon reflects a change in technological practice, rather than a change in source area, with the lower $\delta^{18}O$ of the late Roman and Islamic glasses reflecting dune or kurkar material, rather than beach sand *sensu stricto*, as has been suggested above to account for the differences in ε_{Nd} between glass and sand. If so, we may be seeing a movement away from production located on the beaches to production located up to a few kilometres away from the sea. Such a change might be driven, for example, by the availability of wood for fuel. If such a change in location occurred, the earlier production close to the shore might have been lost to the archaeological record, due to erosion and/or burial.

Oxygen isotope analyses of plant ash glasses revealed $\delta^{18}O$ values significantly lower than for the natron glasses, ranging from 11.8‰ to 12.9‰. We suggested (Leslie *et al.* 2006) that this might have been in part due to the use of glassmaking sands which were not from the Levantine coast, but the Nd isotope study (Degryse *et al.*, forthcoming) indicates that Banias and Tyre glasses are indeed likely to have been made from coastal sands. Hence, the difference in O-isotope compositions between natron and plant ash glasses is likely to reflect significantly lower O-isotope ratios in the plant ashes, relative to those of natron. We are not aware of published O-isotope analyses of plant ashes for comparison; these are likely to differ significantly from those of the dried plant due to the major loss of oxygen in the ashing process. However, it is quite possible that the similar processes giving rise to the $\delta^{18}O$ of plant ash will produce ashes of a generally similar oxygen isotopic composition, and that differences in $\delta^{18}O$ between plant ash glasses reflect differences in the sources of silica. Data are needed for $\delta^{18}O$ in a range of plants and their ashes to investigate this possibility.

The difficulty in the interpretation of oxygen isotope data is that $\delta^{18}O$ for a glass is a reflection of the composition of both the silica source and the flux. In the case of natron glasses, the silica source is likely to provide the majority of the oxygen (assuming that lime, magnesia, potash are supplied by the sand), and it can be assumed that the natron used was predominantly Egyptian, arguably with a relatively narrow range of isotope compositions. However, in the case of plant ash glass the situation is less clear-cut. Furthermore, the range in $\delta^{18}O$ of sands is relatively limited. Thus oxygen can be useful as a discriminating tool, but less effective as a predictor of provenance than neodymium.

Lead isotopes

Few analyses have been carried out to determine the Pb-isotope composition of glass from workshops in the Levant. We have analysed the composition of two glass chunks from Apollonia (Arsuf); the results are presented in Table 1.1 (for analytical method see appendix). Major element compositions of these samples are given by Freestone *et al.* 2000. The isotopic compositions of these glasses are close to those of the four samples of natron glass from Raqqa analysed by Henderson *et al.* (2005) and very probably made close to the Levantine coast (cf. *supra*). It is likely that these compositions provide an indication of typical Pb-isotope compositions of Levantine coastal glass.

	$^{208}Pb/^{206}Pb$	$^{207}Pb/^{206}Pb$	$^{206}Pb/^{204}Pb$	$^{208}Pb/^{204}Pb$	$^{207}Pb/^{204}Pb$
6831-11	2.0814	0.8446	18.541	38.588	15.659
6831-12	2.0777	0.8419	18.602	38.648	15.660

Table 1.1
Lead isotope composition of glass from Apollonia, Israel

The Pb contents of the two glasses from Apollonia are 29 and 30 ppm, but it is not clear which phases in the glassmaking sand contained this lead, so interpretation is not straightforward. Sulphides such as galena are unstable in air and oxygenated seawater and are unlikely to have been present. Thus the lead is likely to have been carried in a mineral phase such as alkali feldspar or possibly adsorbed from seawater into an organic phase such as molluscan shell. HCl soluble lead from Mediterranean sediments has a very similar composition to that in the glasses (Chow 1968: $^{206}Pb/^{204}Pb = 18.42$, $^{207}Pb/^{206}Pb = 0.852$, $^{208}Pb/^{206}Pb = 2.076$), consistent with their origin from Levantine coastal sediments. However, it should be noted that the average composition of the continental crust is also similar isotopically (Chow and Patterson 1962: $^{206}Pb/^{204}Pb = 18.58$, $^{207}Pb/^{204}Pb = 15.77$, $^{208}Pb/^{204}Pb = 38.87$). It is possible that the lead isotopic composition of Levantine beach sand may reflect such an averaging of the sedimentary signature that the isotopes of lead may not be sensitive discriminants between glass production centres in many cases.

Predictive provenancing: HIMT glass

HIMT glass, comprising a group of natron glasses which are high in transition metals (= high iron manganese titanium: Freestone 1994) is arguably the most abundant glass type in the late 4th to 5th centuries AD, but its production site has not been identified. Elemental analysis (Freestone et al. 2005) indicates strong inter-element correlations between Fe, Ti, Mn, Mg and Al, and the work of Foy et al. (2003a, b) indicates the occurrence of similar correlations in the trace elements. Correlations of this type have been widely identified in early glass (e.g. Henderson et al. 2004, Shortland 2005) and may normally be attributed to contamination of a single sand source by heavy minerals or clay. However, such variations normally involve relatively low concentrations of iron, etc. (well under 1%), whereas iron oxide in HIMT may range up to 3% or more.

	Site	$^{208}Pb/^{206}Pb$	$^{207}Pb/^{206}Pb$	$^{206}Pb/^{204}Pb$	$^{208}Pb/^{204}Pb$	$^{207}Pb/^{204}Pb$	$^{87}Sr/^{86}Sr$
33027 W	Carthage	2.0637	0.8353	18.7017	38.5955	15.6220	0.708079
32831 X	Carthage	2.0847	0.8460	18.5028	38.5720	15.6520	0.708858
32832 V	Carthage	2.0706	0.8378	18.6871	38.6928	15.6560	0.708138
32833 T	Carthage	2.0794	0.8391	18.7123	38.9017	15.7038	0.708580
6830-20	N Sinai	2.0790	0.8435	18.5398	38.5133	15.6511	0.708785
6830-23	N Sinai	2.0505	0.8292	18.8687	38.6890	15.6465	0.708251
6830-27	N Sinai	2.0737	0.8399	18.6368	38.6471	15.6524	0.708496
6830-28	N Sinai	2.0805	0.8438	18.5861	38.6681	15.6822	0.708572
6830-21	N Sinai	NA	NA	NA	NA	NA	0.708258
6830-67	N Sinai	NA	NA	NA	NA	NA	0.708388
6830-81	N Sinai	NA	NA	NA	NA	NA	0.708526

Table 1.2
Lead and Strontium isotope data for HIMT glass from Carthage and North Sinai

Strontium and lead isotope analyses of a number HIMT glasses from Carthage and North Sinai were carried out by Freestone et al. (2005) and are presented in full in Table 1.2 (for analytical methods see appendix). These indicate that the strontium and lead isotope composition of the glass is correlated with the elemental variations, for example, with Al_2O_3 and MgO (Fig. 1.7). Furthermore, as the concentration of the transition metals in the glass increases, the $^{87}Sr/^{86}Sr$ decreases, moving away from the seawater value (Fig. 1.7). The relationship between lead and strontium isotopes (Fig. 1.8) indicates that, as the terrestrial component of Sr increases, the lead becomes younger. The simplest explanation for this is that young potassic feldspar, $KAlSi_3O_8$, which is known to accommodate relatively high amounts of lead and strontium, is increasing in the glassmaking sand. This is consistent also with the increasing Al_2O_3 concentration, although the elemental data suggest that mafic minerals (pyroxene or amphibole) also increase with the feldspar. Another possibility may be that the glassmaking sand becomes contaminated with a clay-rich sediment, such as soil, which is richer in Pb and Sr derived from the Nile.

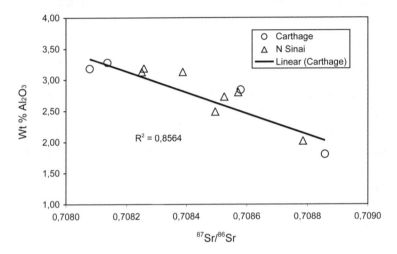

Fig. 1.7
Alumina content versus strontium isotope ratio of HIMT glass from Carthage and the North Sinai survey

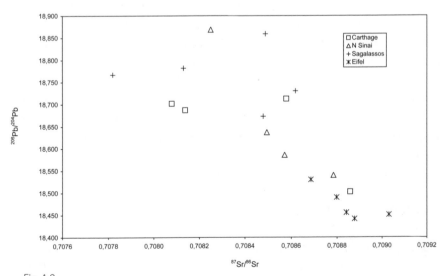

Fig. 1.8
Relationship between lead and strontium isotopes in HIMT glass from Carthage, North Sinai and Sagalassos (Degryse *et al.* 2006). 4th century AD glass from the Eifel region in Germany (Wedepohl and Baumann 2000) is also shown

Lead and strontium isotopes in five samples of HIMT glass from Sagalassos were measured by Degryse *et al*. (2006). These data are shown for comparison in Fig. 1.8. They correspond well to that part of the distribution characterized by younger lead and older strontium, as defined here. Wedepohl and Baumann (2000) analysed 4th century glass from workshops in the Eifel area of Germany (Hambach Forest and Gellep). $^{87}Sr/^{86}Sr$ values approaching those of Holocene seawater were interpreted as indicating the deliberate addition of crushed shell to the batch, while Pb-isotopes were considered to indicate a source of sand in the Eifel region. However, as noted by Freestone *et al*. (2003) and Degryse *et al*. (2006), the isotopic data are somewhat ambiguous with respect to the origin of these glasses. In fact, comparison of the Eifel data with those for HIMT shows that their isotopic compositions are very similar (Fig. 1.8). Furthermore, the high Zr and Ba contents of the Eifel glasses are replicated in HIMT (Freestone *et al*. 2005). A common origin for these glasses must be considered likely.

Neodymium, strontium and lead isotope measurements on HIMT glass, including samples from London, have been reported by Freestone *et al*. (in press). They found that ε_{Nd} lay in the range -6.0 to -5.0, similar to the range in glass from Levantine tank furnaces, suggesting an origin for the glasses in the Levant. They noted that the lower $^{87}Sr/^{86}Sr$ ratios of the HIMT range favoured an area of origin on the Egyptian coast, between Alexandria and Gaza, where the Nile strontium isotope signature dominates the sediments (Weldeab *et al*. 2002). The young feldspar component, inferred from the lead and strontium results above, would thus be derived from the Nile, and ultimately the volcanics of the Ethiopian highlands. This model is also consistent with the lead isotope data for glass from the Levant, which are very similar to those of the particular HIMT glasses which lie at the end of the range with more marine strontium and older lead (Table 1.1). Thus a comprehensive range of isotopic data suggests an Egyptian origin for HIMT glass. By implication, the glasses from Eifel previously considered to represent local manufacture of primary glass are also likely to originate in the south-eastern Mediterranean.

Comparison and discrimination: plant ash glass

The analysis of oxygen and strontium isotopes in plant ash glasses from Tyre and Banias by Leslie *et al*. (2006), plus the analysis of neodymium and strontium by Degryse *et al*. (forthcoming), allows some comment on the discriminating capacities of these isotopes for plant ash glass from the Levantine region. In terms of differences in major element composition, it seems unlikely that the Banias glass was made in the Tyre furnaces (Fig. 1.2; Freestone 2002). Leslie *et al.* (2006, also Freestone 2006) showed, on the basis of a small number of analyses, that Tyre and Banias glasses are separated by $^{87}Sr/^{86}Sr$, indicating the use of plant ash from different areas, supporting the inference of a different production site for the Banias glass. However, there is some overlap in terms of $\delta^{18}O$, similar to the overlap also seen in the Nd isotope signatures from the two sites. This is not surprising, as both Nd and O isotopes are a reflection of sand composition. It appears that the Banias glass was made from a coastal sand similar to that used in the Tyre furnaces, but using a different plant ash. The possibility that the two glass types were both made at Tyre, using a similar sand but at different times, seems unlikely, as the Al_2O_3 content of the Banias glasses is lower than that of Tyre. This suggests that there is likely to have been another furnace complex on the Levantine coast making plant ash glass in the Islamic period.

Plant ash glass from Raqqa in Syria has been analysed for Sr, O and Pb isotopes by Henderson *et al.* (2005). Major elemental compositions are not given and groups not indicated, but the numbering in the paper cited suggests that the plant ash glasses analysed were from compositional types 1 and/or 2 of Henderson *et al.* (2004). These are similar in general respects to the Tyre and Banias glasses, so a comparison of their isotopic compositions is of interest. The Raqqa glass has similar $^{87}Sr/^{86}Sr$ to Tyre but a very different $\delta^{18}O$ from both Tyre and Banias (Leslie *et al.* 2006, Freestone 2006), suggesting that the sand used in the Raqqa glass is unlikely to have originated in the coastal area (although Nd data are needed to confirm this). This is consistent with the proposal of Henderson *et al.* (2005) that the Raqqa plant ash glass was a primary type made at Raqqa (although the archaeological evidence presented so far is not conclusive in this respect). The similarity of the strontium isotope ratios is consistent with the use of plant ash from the same area in both the Raqqa and Tyre glasses. Given that trade in Syrian plant ashes is well known from the late medieval and early modern periods (Ashtor and Cevidalli 1983), this is a real possibility. However, given that we are dealing with a single parameter ($^{87}Sr/^{86}Sr$) with a limited range, the possibility that this overlap in Sr compositions is coincidental cannot be ignored. Even so, it represents potentially the earliest evidence for a trade in plant ashes for glassmaking.

Discussion and conclusion

Based on the limited available data, we have provided an outline of the isotopic composition of Levantine glass, in terms of the isotopes of strontium, neodymium, oxygen and lead. Isotopic analysis appears very promising from the perspective of the investigation of early glass production. In particular Nd would appear to be a strong marker for eastern Mediterranean sands, while Sr determines the relative contribution of shell and/or marine sand to the glassmaking batch in the case of natron glasses, and can discriminate between the use of plant ashes from different sources in the case of plant ash glasses.

The contribution of both flux and sand to the isotopic composition of oxygen makes it somewhat less powerful than neodymium in some respects, but it can be an effective discriminant. Furthermore, unlike neodymium, which reflects the heavy (non-quartz) mineral content of a sand, oxygen isotopes are mainly determined by the quartz itself, which may prove useful in identifying the use of different varieties of quartz such as pebbles and chert in the glass batch. Laser fluorination of oxygen is faster, requires less sample and is less expensive than many other forms of isotopic analysis. Lead is somewhat problematic; in the present investigations, we have been fairly confident that the glass has not been contaminated by the recycling of coloured glasses, but the inclusion of recycled pigments (or, indeed, lead-backed mirror glass) could have a significant effect on isotopic composition. The few data we have suggest that the lead isotope composition of Levantine glass is close to that of lead in Mediterranean sediments, but this is close to the average composition of crustal lead, so its effectiveness as a discriminant is still unclear.

In general it is clear that a combination of isotopes is likely to be more powerful than any one system in the investigation of early glass production. Furthermore, isotopic systems such as those of Pb and O, although likely to be of more restricted application than Nd and Sr, may make a useful contribution in some cases. For example, lead was helpful in the interpretation of HIMT glass, and a good set of O-isotope data for the HIMT dataset might help to clarify further the origins of the strongly correlated elemental compositions in this group by indicating any variation in the source of the quartz.

Overall, we must emphasize the very limited number of data upon which this survey is based. While the range of ε_{Nd} of glasses made in Levantine furnaces, almost certainly from local sand, is currently known to be at least -6.0 to − 3.2, we do not know how much greater this spread will prove to be for the whole population. On the basis of simple statistical considerations, we should not expect to have defined the full range of compositions at this stage. Significantly less positive as well as more positive compositions are likely to occur. As the range of Levantine Nd compositions increases with further analysis, it is possible that ε_{Nd} will prove to be a less powerful interpretive tool than might be hoped, although its strength as a signature of Levantine glass in general is likely to be assured. The analysis of raw materials, as well as more glass from primary furnaces, is essential.

To conclude, while of great potential, isotopic studies are unlikely to prove a panacea in glass provenance studies. Its two particular successes - the use of strontium to recognize a marine component and the use of neodymium to recognize sand derived from the Nile - are extremely useful, but are not universally applicable. The range in isotopic composition of a single glass production centre is rather large relative to the total spread of isotopic compositions of Sr, Nd, O and Pb in natural sands and discrimination of sources is likely to require the measurement of multiple isotopes, in conjunction with elemental compositions.

Acknowledgements

The Renaissance Trust and the late Mr Gerry Martin supported our early isotope work. K.H. Wedepohl and S. Bowman encouraged our interest in isotopic studies of early glass. We thank our many collaborators, most of whom are referenced, but especially Patrick Degryse, Michael Hughes, Sarah Jennings, Karen Leslie and David Lowry.

References

F. Aldsworth, G. Haggarty, S. Jennings, D. Whitehouse, 2002, Medieval glassmaking at Tyre, Lebanon, Journal of Glass Studies, 44, 49-66.

J.W. Arden, N.H. Gale, 1974, New electrochemical technique for the separation of lead at trace levels from natural silicates, Analytical Chemistry, 46, 2-9.

E. Ashtor, G. Cevidalli 1983. Levantine alkali ashes and European industries. European Journal of Economic History 12, 475-522.

Y. Barkoudah, J. Henderson, 2006, Plant ashes from Syria and the manufacture of ancient glass: ethnographic and scientific aspects, Journal of Glass Studies, 48, 297-321.

R.H. Brill, 1970, Lead and oxygen isotopes in ancient objects, Philosophical Transactions of the Royal Society of London, A.269, 143-164.

R.H. Brill, 1988, Scientific investigations, in: G.D. Weinberg, Excavations at Jalame: Site of a Glass Factory in Late Roman Palestine, University of Missouri, 257-294.

R.H. Brill, T.K. Clayton, C.P. Stapleton, 1999, Oxygen isotope analysis of early glasses, in: R.H. Brill, Chemical analyses of early glasses, Corning Museum of Glass, 303-322.

T.J. Chow, 1968, Lead isotopes of the Red Sea region, Earth and Plantery Science Letters, 5, 143-147.

T.J. Chow, C.C. Patterson, 1962, The occurrence and significance of lead isotopes in pelagic sediments, Geochimica et Cosmochimica Acta, 26, 263-308.

P. Degryse, J. Schneider, U. Haack, V. Lauwers, J. Poblome, M. Waelkens, Ph. Muchez, 2006, Evidence for glass 'recycling' using Pb and Sr isotopic ratios and Sr-mixing lines: the case of early Byzantine Sagalassos, Journal of Archaeological Science, 33, 494-501.

P. Degryse, J. Schneider, 2008, Pliny the Elder and Sr-Nd radiogenic isotopes: provenance determination of the mineral raw materials for Roman glass production. Journal of Archaeological Science, 35, 1993-2000.

P. Degryse, I.C. Freestone, J.Schneider, S. Jennings, in press, Technology and provenance study of Levantine plant ash glasses using Sr-Nd isotope analysis, Archaeologische Korrespondezblatt, 2009.

K.O. Emery, C.J. George, 1963, The shores of Lebanon, Miscellaneous Paper in the Natural Sciences. Beirut, American University.

K.O. Emery, D. Neev, 1960, Mediterranean Beaches of Israel, Geological Survey of Israel Bulletin, 26, 1-23.

A. Foucault, D.J. Stanley, 1989, Late quaternary palaeoclimatic oscillations in East Africa recorded by heavy minerals in the Nile Delta, Nature, 339, 44-46.

D. Foy, M.Picon, M Vichy, 2000, Les Matières premières du verre et la question des produits semi-finis, Antiquité et Moyen Âge in: Arts du Feu et Productions Artisanales. XXe Recontres Internationales d'Archéologie et d'Histoire d'Antibes, Éditions APDCA, 419-432

D. Foy, M. Picon, M.Vichy, 2003a, Verres Omeyyades et Abbasides d'origine Egyptienne: Les temoignages de l'archéologie et de l'archéometrie, in: Annales 15e Congrès de l'Association Internationale pour l'Histoire du Verre, 138-143.

D. Foy, M. Picon, M. Vichy, V. Thirion-Merle, 2003b, Caractérisation des verres de la fin de l'Antiquité en Mediterranée occidentale : l'emergence de nouveaux courants commerciaux, in: D.Foy, M.D. Nenna (eds.) Échanges et Commerce du Verre dans le Monde Antique, Éditions Monique Mergoil, 41-85.

I.C. Freestone, 1994, Chemical analysis of «raw» glass fragments, in: H R Hurst, Excavations at Carthage, Vol II, 1 The Circular Harbour, North Side 290, Oxford Univ Press for British Academy.

I.C. Freestone, 2002, Composition and Affinities of Glass from the Furnaces on the Island Site, Tyre, Journal of Glass Studies, 44, 67-77.

I.C. Freestone, 2006, Glass production in Late Antiquity and the Early Islamic period: a geochemical perspective, in: M. Maggetti, B. Messiga (eds) Geomaterials in Cultural Heritage, Geological Society of London Special Publication, 201-216.

I.C. Freestone, Y. Gorin-Rosen, M.J. Hughes, 2000, Primary glass from Israel and the production of glass in late antiquity and the early Islamic period, in : M.-D. Nenna (ed.) La Route du Verre: Ateliers primaires et secondaires de verriers du second millénaire av. J.-C. au Moyen-Âge, Travaux de la Maison de l'Orient Méditérranean 33, 65-83.

I.C. Freestone I. C., Greenwood R. and Gorin-Rosen Y. (2002a) Byzantine and early Islamic glassmaking in the Eastern Mediterranean: production and distribution of primary glass. in G. Kordas (ed) Hyalos - Vitrum - Glass. History Technology and Conservation of glass and vitreous materials in the Hellenic World, 167-174.

I.C. Freestone, M. Ponting, J. Hughes, 2002a, The origins of Byzantine glass from Maroni Petrera, Cyprus, Archaeometry 44, 257-272.

I.C. Freestone, K. A. Leslie, M. Thirlwall, Y. Gorin-Rosen, 2003, Strontium isotopes in the investigation of early glass production: Byzantine and early Islamic glass from the Near East, Archaeometry 45, 19-32.

I.C. Freestone, S. Wolf, M. Thirlwall, 2005, The production of HIMT glass: elemental and isotopic evidence, in: Proceedings of the 16th Congress of the Association Internationale pour l'Histoire du Verre, London, 153-157.

I.C. Freestone, R.-E. Jackson-Tal and O. Tal, 2008a, Raw glass and the production of glass vessels at Late Byzantine Apollonia Arsuf, Israel, Journal of Glass Studies 50, 67-80.

I.C. Freestone, M.J. Hughes, C.P.Stapleton, 2008b, The composition and production of Anglo-Saxon glass, in: Evison VI, Catalogue of Anglo-Saxon Glass in the British Museum, 29-46.

I.C. Freestone, P. Degryse, J. Shepherd, Y. Gorin-Rosen, J. Schneider, in press, Neodymium and Strontium Isotopes Indicate a Near Eastern Origin for Late Roman Glass in London, Journal of Archaeological Science

N.H. Gale, 1996, A new method for extracting and purifying lead from difficult matrices for isotopic analysis, Analytica Chimica Acta, 332, 15-21.

H. Gerstenberger, G. Haase, 1997, A highly effective emitter substance for mass spectrometric Pb isotope ratio determinations, Chemical Geology, 136, 309-312.

Y. Gorin-Rosen, 1995, Hadera, Bet Eli'ezer, Excavations and Surveys in Israel, 13, 42-43.

Y. Gorin-Rosen, 2000, The ancient glass industry in Israel: summary of new finds and new discoveries, in: M.D. Nenna (ed.) La Route du Verre: Ateliers primaires et secondaires de verriers du second millénaire av. J.-C. au Moyen-Âge, Travaux de la Maison de l'Orient Méditérranean, 33, 49-64.

B. Gratuze, J.N. Barrandon, 1990, Islamic glass weights and stamps: analysis using nuclear techniques, Archaeometry, 32, 155-162.

J. Henderson, S.D. McLoughlin, D.S. McPhail, 2004, Radical changes in Islamic glass technology: evidence for conservatism and experimentation with new glass recipes from early and middle Islamic Raqqa, Syria, Archaeometry, 46, 439-468.

J. Henderson, J. Evans, H. Sloane, M. Leng, C. Doherty, 2005, The use of strontium, oxygen and lead isotopes to provenance ancient glasses in the Middle East, Journal of Archaeological Science, 32, 665-674.

M.D. Krom, D.J. Stanley, R.A. Cliff, J.C. Woodward, 2002, Nile River sediment fluctuations over the past 7000 yr and their key role in sapropel development, Geology, 30, 71-74.

K.A. Leslie, I.C. Freestone, D.Lowry, M. Thirlwall, 2006, The provenance and technology of Near Eastern glass: oxygen isotopes by laser fluorination as a complement to strontium, Archaeometry, 48, 253-270

M.D. Nenna, 2007, La Production et la circulation du verre au Proche-Orient : État de la question, Topoi soppl., 8, 123-150.

M.D. Nenna, M. Vichy, M. Picon, 1997, L'Atelier de verrier de Lyon, du Ier siècle après J.-C., et l'origine des verres 'Romains', Revue d'Archèomètrie, 21, 81-87.

M.D. Nenna, M. Picon, M Vichy, 2000, Ateliers primaire et secondaires en Égypt à l'époque gréco-romaine, in: M.D. Nenna (ed.) La Route du Verre: Ateliers primaires et secondaires de verriers du second millénaire av. J.-C. au Moyen-Âge, Travaux de la Maison de l'Orient Méditérranean, 33, 97-112.

M.D. Nenna, M Picon, V. Thirion-Merle, M. Vichy, 2005, Ateliers Primaires du Wadi Natrun: nouvelles découvertes, Annales 16e Congrès de l'Association Internationale pour l'Histoire du Verre, 59-63.

Y. Nir, 1996, The city of Tyre, Lebanon and its semi-artificial tombolo, Geoarchaeology, 11, 235-250.

M. Picon, M. Vichy, 2003, D'Orient et Occident: l'origine du verre à l'époque romaine et durant le haut Moyen Age, in: D. Foy, M.D. Nenna (eds.) Échanges et Commerce du Verre dans le Monde Antique, Éditions Monique Mergoil, 17-32.

M. Pomerancblum, 1966,. The distribution of heavy minerals and their hydraulic equivalents in sediments of the Mediterranean continental shelf of Israel, Journal of Sedimentary Petrology, 36, 162-174.

A.J. Shortland, 2004, Evaporites of the Wadi Natrun: seasonal and annual variation and its implication for ancient exploitation, Archaeometry, 46, 497-516.

A.J. Shortland, L. Schachner, I.C. Freestone, M. Tite, 2006, Natron as a flux in the early vitreous materials industry –sources, beginnings and reasons for decline, Journal of Archaeological Science, 33, 521-530.

A.J. Shortland, 2005, The raw materials of early glasses: the implications of new LA-ICPMS analyses, Annales 16e Congrès de l'Association Internationale pour l'Histoire du Verre, 1-5.

D.J. Stanley, J.G. Wingerath, 1996, Clay mineral distributions to interpret Nile cell provenance and dispersal: 1. Lower River Nile to Delta Sector, Journal of Coastal Research, 12, 911-929.

Z. Stos-Gale, N.H. Gale, J. Houghton, R. Speakman, 1995, Lead Isotope data from the Isotrace Laboratory, Oxford: Archaeometry database 1, ores from the western Mediterranean, Archaeometry, 37, 407-415.

K. Tachikawa, M. Roy-Barman, A. Michard, D. Thouron, D. Yeghicheyan, C. Jeandel, 2004, Neodymium isotopes in the Mediterranean Sea: comparison between seawater and sediment signals. Geochimica et Cosmochimica Acta 68, 3095-3106.

O. Tal, T.E. Jackson-Tal, I.C. Freestone, 2004, New Evidence of the Production of Raw Glass at Late Byzantine Apollonia-Arsuf (Israel), Journal of Glass Studies, 46, 51-66.

M. Valloto, M. Verità, 2002, Glasses from Pompeii and Herculaneum and the sands of the rivers Belus and Volturno, in: J. Renn, G. Castagnetti (eds.) Homo Faber: Studies on Nature, Technology and Science at the Time of Pompeii, L'Erma di Bretschneider, 63-73.

K.H. Wedepohl, A. Baumann, 2000, The use of marine molluskan shells for Roman glass and local raw glass production in the Eifel area (Western Germany), Naturwissenschaften, 87, 129-132.

S. Weldeab, K.C. Emeis, C. Hemleben, W. Siebel, 2002, Provenance of lithogenic surface sediments and pathways of riverine suspended matter in the Eastern Mediterranean Sea: evidence from [143]Nd/[144]Nd and [87]Sr/[86]Sr, Chemical Geology, 186, 139-149.

Appendix: analytical methods

These refer only to the previously unpublished data tabulated in the paper. For other data the original publications should be consulted.

DETERMINATION OF ISOTOPIC COMPOSITION OF LEAD

Lead isotope analysis was carried out by S. Wolf at the Isotrace Laboratory, Research Laboratory for Archaeology and the History of Art, Oxford University.

The surface of each glass sample was carefully abraded with abrasive paper and the samples cleaned with deionized water in an ultrasonic bath for 5 minutes. Two milligrams of the dried glass samples were then weighed and ground to fine powder in an agate mortar. Lead separation was performed inside a better than Class 100 chemical workstation inside a Class 1000 overpressured clean room, following a strict clean laboratory protocol (Arden and Gale 1974). Samples were digested in HF and HCl, evaporated to dryness and redissolved in dilute HCl. The solutions were run through cation exchange columns, and the dried residues redissolved in weak nitric acid (HNO_3). The lead was deposited electrochemically and the lead fraction recovered after separation was dissolved in dilute HNO_3 and loaded with 1 M H_3PO_4 and an emitter, comprising a mixture of colloidal silicic acid and dilute phosphoric acid (Gerstenberger and Haase, 1997), onto a previously outgassed Rhenium filament.

Lead isotope measurements were carried out using a VG 38-54-30 thermal ionisation mass spectrometer. Data were acquired in the static four collector mode with ^{208}Pb ion beam intensities between 0.5×10^{-11} and 2×10^{-11} A. Each run consisted of 60 measurements of three independent lead isotope ratios. Standard errors were generally below 0.05%. The data were corrected for fractionation in the mass spectrometer by 1.001 per a.m.u. with reference to 53 measurements of the NIST/NBS lead isotope standard SRM981. More comprehensive descriptions of the separation method can be found in Gale (1996) and of the measurements by Stos-Gale et al. (1995).

DETERMINATION OF ISOTOPIC COMPOSITION OF STRONTIUM

Measurements were conducted in the Department of Geology, Royal Holloway University of London. Fragments were carefully cleaned to remove contaminants and digested in nitric and hydrofluoric acids, evaporated to dryness then converted to chlorides using a two-step acidification with HNO_3 and HCl. Solutions in 2.5M HCl were run through conventional cation exchange columns, the Sr-bearing fractions collected and dried onto degassed single tantalum filament beads using phosphoric acid. The beads were analysed for strontium isotopes using a multicollector VG354 thermal ionisation mass spectrometer. Accuracy was monitored using NBS standard SRM 987, which yielded $^{87}Sr/^{86}Sr = 0.710248 \pm 0.000015$ (2σ on all analysed standards, n>130) over the year in which the samples were analysed.

Neodymium and strontium isotopes in the provenance determination of primary natron glass production

Patrick Degryse, Jens Schneider, Veerle Lauwers, Julian Henderson, Bernard Van Daele, Marleen Martens, Hans (D.J.) Huisman, David De Muynck, Philippe Muchez

Introduction

The great majority of ancient glass was chemically based upon silica fluxed with soda or potash. The earliest known glass was found in Late Bronze Age Mesopotamia and Egypt. It was a soda-lime silica glass, and this type predominated across Western Asia and the Mediterranean right up to the modern period (Freestone 2006). Chemically, ancient soda-lime-silica glass falls into two categories (Sayre and Smith 1961): (1) plant ash glass, combining a plant ash with quartz pebbles, and (2) natron glass, combining soda-rich mineral matter with quartz sand. Natron glass was the predominant type of ancient glass in the Mediterranean and Europe from the middle of the first millennium BC until the 9th century AD (Henderson 1989, Freestone *et al.* 2002a, Henderson 2003, Shortland 2004). Work by Foy *et al.* (2003) suggests that there are likely to have been around 10 major glass groups in the Mediterranean and Western European region between the 1st and 9th centuries AD. Before that time plant ash glass was also produced, mainly in Egypt and Mesopotamia. Throughout the Mediterranean and Europe, however, using plant ashes as a flux became dominant practice only from the 9th century onwards (Henderson *et al.* 2004, Freestone 2006).

Initially it was assumed that glass was made in the same workshops where the vessels, windows etc. were being formed. However, the discovery of raw glass in the form of ingots in the Late Bronze Age (Nicholson *et al.* 1997, Rehren and Pusch 1997) and as lumps of glass (chunks) in the Roman and early medieval periods (Foy *et al.* 2000) suggests the export of glass chunks as an economic commodity. *Primary* workshops which made the raw glass were, in many cases, clearly distinct from the *secondary* workshops which shaped the glass vessels. A single primary workshop could then supply many secondary workshops over a large geographical area (Gorin-Rosen 2000, Nenna *et al.* 2000). However, it is necessary still to be cautious about applying such models too rigidly to ancient economies, and to some extent the separation of glassmaking and glassworking activities may have been dependant on the scale of production. For example, in an (inland) urban environment why would primary and secondary glass production necessarily be separated physically by any great distance? If a glassblowing workshop was set up close to a primary glassmaking furnace, then clearly raw glass could be supplied directly to the local glassblowers. Indeed evidence of this was discovered during the excavation of a 9th century AD Islamic extra-mural industrial complex at al-Raqqa, northern Syria (Henderson *et al.* 2005a). This is not to say that raw glass manufactured at al-Raqqa was not also exported to other glassworking centres. For late Roman glass production, theories are centred on two models (Jackson *et al.* 2003). The first states that contemporary natron glass production was divided between a relatively small number of workshops which made raw glass and a large number of secondary workshops which fabricated vessels (Freestone 2006). It is clear from excavation that large quantities of natron glass were being made from its mineral raw materials in a relatively limited number of primary glass production centres mainly in Egypt in the 1st to 3rd centuries AD and Syro-Palestine in the 4th to 8th centuries AD (Brill 1988, 1999, Freestone *et al.* 2000, 2002a, Picon and Vichy 2003). Suggestions that similar units existed in the Levant in early Roman times have only recently been proven, with the discovery of early Roman primary glassmaking furnaces in Beirut, Lebanon (Kouwatli *et al.* 2008). It has been argued that Roman blue-green glass and later glass produced in the Levant are sufficiently similar for it to be likely that Roman glass was made there (Nenna *et al.* 1997, Picon and Vichy 2003, Foy *et al.* 2003), although archaeological and scienctific evidence is difficult to interpret (Baxter *et al.* 2005). Some authors have suggested that early Roman primary production may have taken place elsewhere (Leslie *et al.* 2006, Jackson *et al.* 2003). Moreover, the second model of late Roman glass production proposes the existence of local glassmaking and -working centres (Wedepohl *et al.*, 2003). Also, there is evidence which supports the manufacture of primary glass in Roman Europe. The ancient author, Pliny the Elder, writing before 79 AD, indicates in his Natural History (Hist. Nat XXXVI, 194) that sands from the coast of Italy between Cumae and Literno near Naples and the 'Spanish and Gaulish provinces' were also used (Freestone *et al.* in press). This, however, has never been confirmed by excavations, although the suitability of some of the sands explicitly described by Pliny has been suggested (Silvestri *et al.* 2006).

The concept of a *division of production* leads to a very different interpretation of analytical data, so that glass compositions reflect predominantly the primary glassmaking sources, rather than the secondary workshops in which the objects were made (e.g. Nenna *et al.* 1997; Foy *et al.* 2000; Freestone *et al.* 2000, 2002b). This model has significant implications for the study of ancient glass production based upon the chemical analysis of glass artefacts. While for several decades clay-based ceramics have been routinely subjected to elemental analysis to determine provenance, the application of these methods to archaeological glass has thus proved far less tractable (Freestone 2006). The combined effects of the mixing of primary resources and the recycling of glass can stymie attempts to identify the origin of glass raw materials based upon elemental analysis alone.

Glass provenancing

GLASS PROVENANCING AND ELEMENTAL ANALYSIS

A great deal of effort has gone into the major element analysis of glass (Brill 1999). For the most part, this has not led to meaningful groupings with respect to the geographical origin of the mineral resources. For example, all Roman glass was found to be relatively homogeneous natron glass with little variation in major element composition (Freestone 2006). Though significant advances have been made, progress toward an understanding of the exploitation of raw materials, technology and trade through main and trace element analysis remains limited (Freestone 2006). Most progress has been made in studying trace elements like lime, iron, magnesium and alumina, as they can be related to the concentrations of specific minerals (feldspars, micas and clays) in the glassmaking sand. Trace elements in glass have been exploited to separate compositional groups, and the implication has been made that individual objects with these trace element signatures were produced from the same 'batch' (Freestone 2006). However, the presence of elevated transition metals has indicated that scrap glass, including small quantities of coloured glass, may have been incorporated into a batch, pointing to 'recycling' material, and this complicates the picture (Henderson 1993, Jackson 1997). Studies by Freestone *et al.* (2000, 2002b) and Aerts *et al.* (2003) have used trace elements as specific indicators of the origin of glass raw materials. Huisman *et al.* (in press) used trace element composition to source decolorants (Sb) used in the production of roman colourless glass.

GLASS PROVENANCING AND ISOTOPES

Recent studies (Wedepohl and Baumann 2000, Freestone *et al.* 2003, Henderson *et al.* 2005, Degryse and Schneider 2008, Degryse *et al.* 2005, 2006a and b, Freestone *et al.* in press, Henderson *et al.* in press) have shown that the use of radiogenic isotope systems, specifically for strontium (Sr) and neodymium (Nd), has led to the development of new approaches in the provenancing of primary glass, even after its transformation or recycling in secondary workshops.

Sr in ancient glass is mainly incorporated with the lime-bearing material, being shell, limestone or plant ash (Wedepohl and Baumann 2000). Where the lime in glass was derived from Holocene sea shell, the Sr isotopic composition of the glass reflects that of modern seawater (Wedepohl and Baumann 2000). Where the lime was derived from 'geologically aged' limestone, the signature of the glass reflects that of the limestone, possibly modified by diagenesis (Freestone *et al.* 2003, Henderson *et al.* 2005).

Nd in glass is likely to have originated from the heavy mineral content of the sand raw material used. Rare earth element (REE) patterns have been suggested before as a means of distinguishing sand raw material sources (Freestone *et al.* 2002b). Nd isotopes are used as an indicator of the provenance of detrital sediments in a range of sedimentary basin types (Banner 2004). The Nd isotopic composition of the earth's crust shows a wide variation, from ε_{Nd} -45 to +12, but sediments tend to be homogenized so that the sedimentary loads of most of the world's major rivers and airborne dusts vary between ε_{Nd} of -16 and -3 (Goldstein *et al.* 1984). This is the range within which many glassmaking sands are likely to fall (Freestone *et al.* in press). Due to its geological and geographical variability Nd offers great potential in tracing the origins of primary glass production in ancient times. Moreover, the effect of recycling on the Nd composition of a glass batch does not seem to be significant (since there are no high Nd glasses which could modify the base composition of the glass, unlike e.g. lead isotopes affected by high lead glasses in re-melting), nor is the effect of adding colorants or opacifiers (Freestone *et al.* 2005).

Though largely unexplored, Nd isotopes show great promise for addressing hypotheses regarding the primary production of glass in the Roman-Byzantine world. A first example is given in the provenance determination of 4th to 8th century AD glass from Syro-Palestine and Egypt (Freestone *et al.* in press). Levantine-type glass of that era has a Nile-dominated Mediterranean $^{143}Nd/^{144}Nd$ signature, lower Nd content, and a high $^{87}Sr/^{86}Sr$ signature close to the Holocene seawater composition. Contemporary HIMT-type glass is made up of a mixture of a Levantine-type glass and an end member with a Nile-dominated Mediterranean $^{143}Nd/^{144}Nd$ signature, higher Nd content and a low $^{87}Sr/^{86}Sr$ signature. The similarity of Levantine and HIMT glass in terms of $^{143}Nd/^{144}Nd$ signature (values between $\varepsilon =$ -6.0 to –5.1), and the fact that these values are similar to Nile-dominated sediments (Weldeab *et al.* 2002, Stanley *et al.* 2003), strongly suggest that HIMT glass comes from an area extending from the Nile delta northwards to the Levant (Freestone *et al.* in press). A second study investigated the primary provenance of 1st to 3rd century AD natron vessel glass (Degryse and Schneider 2008). Different sand raw materials used for primary glass production in this period were distinguished and characterized by combined Sr and Nd isotopic analyses. Again, a Nile-dominated eastern Mediterranean Nd signature (higher than -6.0 ε Nd) characterized some glass, but a different Nd signature (lower than -7.0 ε Nd) was determined for a large number of samples, suggesting a primary production location in the western Mediterranean or north-western Europe. In this way, strontium and neodymium isotopes proved that Pliny's writings were correct: primary glass production was not exclusive to the Levant or Egypt in early Roman days: other factories of raw glass, probably in the Western Roman Empire were in play.

In this study, the primary provenance of Roman-Byzantine natron vessel glass from different sites in the eastern and western Roman Empire is investigated from the perspective of main elemental versus isotopic analysis. These isotope data obtained from the glass samples are compared with the main element data and the known signatures of primary production centres in the eastern Mediterranean.

Methodology

Sampling

Samples were obtained from several locations in the Roman Empire through cooperation with the VIOE (Vlaams Instituut voor Onroerend Erfgoed - excavation at Tienen), the Rijksdienst voor Oudheidkundig Bodemonderzoek (the Netherlands – excavations at Bocholtz and Maastricht), the excavation at Kelemantia (Slovakia) and at Sagalassos (Turkey). A series of 47 glass samples were selected for both main element and Sr and Nd isotope analysis. Most samples represent free-blown vessel glass, but pressed or slumped plates were also analysed; various colours were selected by eye. Sample dates were determined by stratigraphical association.

Chemical analysis

For main element analysis, samples were fused with a $LiBO_2$ flux and then dissolved in 1N HNO_3. Silicon, aluminium, iron, magnesium, calcium, titanium and phosphorus were determined by atomic emission spectrometry (AES) on a Spectrojet III spectrometer. Sodium and potassium contents were obtained from the same solutions by atomic absorption spectrometry (AAS) on a Varian Techtron AA6 spectrometer. Accuracy for both AAS and AES is better than 2%. Analytical precision at the 95% confidence level determined by replicate analysis was better than 0.5%. Detection limits were at the ppm level for both AAS and AES, but concentrations were expressed at the 0.01 % level. Data accuracy was evaluated by analysis of the international standards Basalt BN-01 and GSJ-JB-1, Granite GN-02 and NIM-G, Lujavrite NIM-L, Feldspar NBS-99a and Gabbro MRG-1.

For isotope analysis, samples were weighed into Teflon screw-top beakers and dissolved in a 3:1 mixture of 22 M HF and 14 M HNO_3 on a hot plate. Solutions were dried and dissolved in aqua regia. Aliquots of these solutions were spiked with a highly enriched ^{84}Sr and ^{150}Nd tracer for separate concentration analyses by isotope dilution, whereas unspiked aliquots were used for determination of isotope ratios. For separation of Sr and Nd from the same sample solutions sequential extraction methods developed by Pin *et al.* (1994) were utilized and slightly modified. Sr and REE were separated using 2 M HNO_3 using coupled miniaturized Teflon columns containing 50 μl of EICHROM Sr and TRU resin, respectively, and eluted with deionized H_2O. For separation of Nd, the REE cut was further passed through a column containing 2 ml EICHROM Ln resin. For this, the column was washed with 5.5 ml 0.25 M HCl after adding the sample. Nd was then stripped off using 4 ml 0.25 M HCl. All measurements were performed on a six-collector FINNIGAN MAT 262 thermal ionization mass spectrometer (TIMS) running in static multicollection mode. Sr isotopic ratios were normalized to $^{86}Sr/^{88}Sr = 0.1194$, Nd isotopic ratios were normalized to $^{146}Nd/^{144}Nd = 0.7219$. Repeated static measurements of the NBS 987 standard over the duration of the study yielded an average $^{87}Sr/^{86}Sr$ ratio of 0.71025 ± 0.00002 (2σ, n=22). Repeated measurements of the La Jolla Nd standard yielded $^{143}Nd/^{144}Nd = 0.511848 \pm 0.000009$ (2σ, n = 8). Total procedural blanks (n=6) did not exceed 30 pg Sr and 50 pg Nd and were found to be negligible.

Archaeological context

Sagalassos

It has already been suggested that early Byzantine (6th-7th century AD) blue raw glass from Sagalassos was imported from several production sites in the Levant (Degryse *et al.* 2005, 2006a), whereas HIMT raw glass from Sagalassos corresponded very well to previously described material (Freestone *et al.* 2005), of which the primary production site is placed in Egypt (Freestone *et al.* in press). Conversely, early to late Roman glass at Sagalassos shows a distinctive major element composition, suggesting a different raw material mixture and possible different origin (Degryse *et al.* 2006b). The samples are representative of the common colour varieties of window and free-blown vessel glass from Sagalassos. The chronology was determined by stratigraphical association with Sagalassos red slip ware (Poblome 1999) and Sagalassos common wares (Degeest 2000). Glass from three distinct periods was sampled: imperial (1st-3rd century AD), late Roman (4th-first half 5th century AD) and early Byzantine (second half 5th to 7th century AD).

MAASTRICHT

Sample Ma3a was retrieved from a grave in the Scharnweg in 1986. Besides pottery, several glass objects such as beakers and bowls were recovered. Typochronologically all the material can be dated to the first half of the third century AD (Panhuysen and Dijkman 1987, p.212 and afb.11). Sample M5a was excavated in 1983 under the Hotel Derlon, in layers assigned to the second quarter of the 5th century AD (Dijkman 1993, Fig. 9-C1 and D8).

KELEMANTIA

The Roman auxiliary fort of Iža (Kelemantia) in Slovakia is situated about 4 km east of the confluence of the rivers Waag and Danube. A double ditch was uncovered, together with the remains of more than eleven barracks. The remains of the earth-and-timber fortification all belong to one single construction phase dating between 175 and 179 AD. This secured dating was possible thanks to the discovery of several coins and *terra sigillata* pottery. Comparing the data obtained with historical texts made it possible to link the fort of Kelemantia with the Marcomannic Wars, waged between the Germanic Marcomanni and Quadi and emperor Marcus Aurelius' troops. The wooden construction was laid to waste by German attackers in 179 AD or was dismantled, abandoned and set on fire by the Roman forces themselves when they left. A few years later, under Emperor Commodus' rule, a stone *castellum* was built on exactly the same spot. This stone camp was occupied until the end of the reign of Valentinianus I in 375 AD, when that emperor died at Brigetio and the barbarians invaded the frontier zone. During the excavations substantial amounts of all kinds of material were found, including many glass fragments belonging to different kinds of glass objects like bottles, bowls, windows and pearls. Samples KEL 82, KEL 229, KEL 299 and KEL 300 belong to excavation layers of the earth-and-timber camp, dated to 175-179 AD. Sample KEL 234 comes from excavation layers in the *castellum* and was dated to the 3rd century AD.

BOCHOLTZ

In 2003, a stone sarcophagus was found in Bocholtz (the Netherlands) as a part of an underground burial chamber, close to a known Roman villa (de Groot 2006). The chamber was dated to the last quarter of the 2nd to the first quarter of the 3rd century AD. Glass grave goods were identified and sampled for analysis. Sample BO 106 is free-blown colourless plate with a greenish tinge, Isings type 42b (Isings 1957), dated to the 2nd century AD. Sample BO 109 is a free-blown colourless cylindrical bottle, Isings type 51b (Isings 1957), dated to the late 2nd – early 3rd century AD. Sample BO 123 is a colourless cast or slumped small bowl.

TIENEN

The small Roman town of Tienen is situated in Belgium, and was part of the Roman civitas Tungrorum. It was founded during the reign of Claudius on the road from Cologne to Boulogne. Large-scale excavations in the periphery of the town revealed numerous pottery kilns, traces of iron workings and bronze casting and glass production. In 2001, a glass furnace dated to the 2nd century AD was excavated there (Cosyns and Martens 2002–2003). All samples analysed here belong to this context. They are samples of free-blown blue (aqua) vessel glass. Determinable pieces are fragments of Isings type 50 (square bottles) or Isings type 3 (ribbed bowls).

Results

The analytical data from this study are given in Table 2.1. Sr-Nd isotopic results for the 1st–3rd century glass are taken from Degryse and Schneider (2008). Major element analyses are expressed as weight %; Sr and Nd isotopic compositions are expressed as ratios. The ratio $^{143}Nd/^{144}Nd$ is also expressed as ε Nd, a parameter which indicates the isotopic composition of the sample, relative to a theoretical primordial composition.

All glass can be characterized as low-magnesia, soda-lime-silica glasses (Henderson 2000). All samples can be identified as natron glass. However, samples KEL2 and SAG573 have elevated MgO, K_2O and P_2O_5 contents. This suggests that this early Roman sample is not a pure natron-based glass, but that plant ashes may have been used as a flux, or mixed with natron glass. However, the high Al_2O_3 and Na_2O contents of this sample are not in concordance with 'standard' compositions of such glass. The blue and green glass in this study is naturally coloured by the presence of Fe_2O_3, the colourless glass is decoloured with Sb (e.g. Degryse *et al*. 2005).

The Sr isotopic signature of most of the 1st–3rd century, 4th-5th century and 6th-7th century AD natron glass shows a composition near to that of the modern-day seawater (between 0.7087 and 0.7091 for $^{87}Sr/^{86}Sr$). Some of the 1st–3rd century AD glass has a significantly lower $^{87}Sr/^{86}Sr$ composition (between 0.7075 and 0.7086), while one 1st–3rd century AD glass sample has a significantly higher Sr signature (0.7096) and one 4th–5th century AD sample has an entirely different, much higher Sr isotopic signature (0.7255). The 6th–7th century AD HIMT glass has a lower Sr isotopic signature (between 0.7078 and 0.7085), as already reported by Freestone *et al*. (2005). The plant ash glass sample SAG 573 has a Sr isotopic composition of 0.7086 for $^{87}Sr/^{86}Sr$, while the plant ash glass sample KEL 2 has a Sr isotopic composition of 0.7090.

The Nd isotopic data of the 1st–3rd and 4th–5th century AD natron glass show a wide range in composition, varying between 0.512511 and 0.511974 for $^{143}Nd/^{144}Nd$, between -2.5 and −13.0 for ε Nd. The plant ash glass sample SAG 573 has an Nd isotopic composition of 0.51229 for $^{143}Nd/^{144}Nd$, -6.7 for ε Nd, while the plant ash glass sample KEL 2 has an Nd isotopic composition of 0.51226 for $^{143}Nd/^{144}Nd$, -7.3 for ε Nd. The Nd isotopic data of the 6th–7th century AD natron glass vary much less, between 0.512408 and 0.512345 for $^{143}Nd/^{144}Nd$, between -4.5 and −5.7 for ε Nd, with one exceptional sample of 0.512180 for $^{143}Nd/^{144}Nd$, -8.9 for ε Nd.

Discussion

The blue (aqua) and green glass analysed has not been deliberately coloured or opacified, thus there has been no contamination of the primary raw materials of the base glasses with materials from other sources. The decoloriser in colourless glass is Sb. It is unlikely, however, that this constituent would contribute significantly to the Sr-Nd balance of the glass. All glass analysed was imported to the respective site either as raw glass from primary production centres located outside the territory of the town (e.g. Tienen, Sagalassos) or as finished objects (possible for all sites). The spread in major element composition of the natron glass suggests that different silica raw materials may have been used for several individuals (Fig. 2.1).

Sample	Date	Colour	$^{143}Nd/$ ^{144}Nd	2 s	e Nd	$^{87}Sr/^{86}Sr$
1st-3rd AD						
Maastricht						
Ma 3 a	first half 3rd AD	blue	0,512343	0,000013	-5,7	0,70913
Tienen						
Tie 11	2nd AD	blue	0,512511	0,000009	-2,5	0,70893
Tie 12	2nd AD	blue	0,512267	0,000009	-7,2	0,70899
Tie 17	2nd AD	blue	0,512378	0,000010	-5,1	0,70902
Tie 24	2nd AD	blue	0,512376	0,000013	-5,1	0,70902
Tie 35	2nd AD	blue	0,512219	0,000009	-8,2	0,70886
Tie 37	2nd AD	blue	0,512083	0,000006	-10,8	0,70891
Tie 41	2nd AD	blue	0,512337	0,000009	-5,9	0,70901
Tie 45	2nd AD	blue	0,512174	0,000008	-9,1	0,70904
Tie 48	2nd AD	blue	0,512262	0,000005	-7,3	0,70896
Tie 49	2nd AD	blue	0,512249	0,000010	-7,6	0,70759
Tie 50	2nd AD	blue	0,512362	0,000008	-5,4	0,70898
Bocholtz						
Bo 106	last quarter 2nd AD	colourless	0,512296	0,000008	-6,7	0,70905
Bo 109	late 2nd - early 3rd AD	colourless	0,512298	0,000008	-6,6	0,70903
Bo 123	late 2nd - early 3rd AD	colourless	0,512291	0,000009	-6,8	0,70906
Kelemantia						
Kel 1 - 82/91	175-179 AD	colourless	0,512325	0,000010	-6,1	0,70904
Kel 2 - 229/06	175-179 AD	blue	0,512266	0,000012	-7,3	0,70901
Kel 3 - 229/88	175-179 AD	green	0,512325	0,000009	-6,1	0,70877
Kel 4 - 234/88	175-179 AD	colourless	0,512177	0,000011	-9,0	0,70966
Kel 5 - 300/06	3rd AD	colourless	0,512336	0,000010	-5,9	0,70894
Sagalassos						
Sag 575	1st-3rd AD	blue	0,512430	0,000002	-4,0	0,70894
Sag 717	1st-3rd AD	blue	0,512410	0,000002	-4,4	0,70879
Sag 718	1st-3rd AD	blue	0,512291	0,000005	-6,8	0,70882
Sag 574	1st-3rd AD	colourless	0,512460	0,000002	-3,4	0,70905
Sag 709	1st-3rd AD	colourless	0,512308	0,000006	-6,4	0,70910
Sag 573	1st-3rd AD	green	0,512294	0,000005	-6,7	0,70865
Sag 722	1st-3rd AD	green	nd	nd	nd	0,70886
Sag 721	1st-3rd AD	green	0,512392	0,000013	-4,8	0,70880
Sag 723	1st-3rd AD	green	0,512425	0,000007	-4,1	0,70857
Sag 724	1st-3rd AD	green	0,512374	0,000006	-5,1	0,70901
4th-5th AD						
Sagalassos						
Sag 579	4th-5th AD	colourless	0,512352	0,000002	-5,6	0,70907
Sag 580	4th-5th AD	colourless	0,511974	0,000002	-13,0	0,72548
Sag 713	4th-5th AD	colourless	0,512387	0,000011	-4,9	0,70881
5th-7th AD						
Sagalassos						
Sag H54	5th-7th AD	blue	0,512408	0,000002	-4,5	0,70895
Sag 714	5th-7th AD	blue	0,512406	0,000002	-4,5	0,70887
Sag 583	5th-7th AD	blue	0,512385	0,000009	-5,0	0,70881
Sag 589	5th-7th AD	blue	0,512381	0,000009	-5,0	0,70896
SA04VL8A	5th-7th AD	blue	0,512345	0,000009	-5,7	0,70886
JP 16	5th-7th AD	Co-blue	0,512383	0,000007	-5,0	0,70889
JP 28	5th-7th AD	Co-blue	0,512382	0,000009	-5,0	0,70908
Sag 588	5th-7th AD	colourless	0,512420	0,000002	-4,3	0,70895
SA04VL8B	5th-7th AD	purple	0,512389	0,000005	-4,9	0,70874
Sag 586	5th-7th AD	HIMT	0,512373	0,000009	-5,2	0,70849
SA00JP25B	5th-7th AD	HIMT	0,512355	0,000009	-5,6	0,70782
SA04VL4	5th-7th AD	HIMT	0,512353	0,000009	-5,6	0,70848
Maastricht						
Ma 5 b	sec quarter 5th AD	blue	0,512180	0,000009	-8,9	0,70876

Table 2.1
Analytical data of the glass studied (nd: not determined)

2 s	SiO$_2$	Al$_2$O$_3$	FeO	Na$_2$O	K$_2$O	CaO	MgO	MnO	TiO$_2$	P$_2$O$_5$	Total
	%	%	%	%	%	%	%	%	%	%	%
0,00001	66,20	2,39	0,94	18,75	0,58	6,34	1,17	1,25	0,25	0,01	97,88
0,00001	69,17	2,60	0,26	18,93	0,53	8,41	0,52	0,42	0,02	0,17	101,01
0,00002	68,72	2,85	0,15	17,28	0,53	9,35	0,58	0,41	0,04	0,17	100,07
0,00001	70,17	2,48	0,05	17,55	0,66	7,29	0,45	0,42	0,06	0,20	99,31
0,00002	71,16	1,88	0,51	19,45	0,51	5,95	0,42	0,01	0,05	0,05	99,98
0,00001	69,65	2,68	0,43	17,24	0,72	8,30	0,56	0,98	0,04	0,25	100,84
0,00002	70,06	2,97	0,26	12,33	0,78	8,83	0,75	0,24	0,06	0,25	96,50
0,00001	69,95	2,90	0,26	16,56	0,71	8,12	0,51	0,40	0,04	0,21	99,65
0,00001	69,36	2,68	0,51	19,42	0,57	6,79	0,52	0,25	0,08	0,14	100,31
0,00001	69,23	2,92	0,05	16,45	0,63	8,80	0,49	0,36	0,04	0,20	99,16
0,00001	66,91	3,41	0,05	15,32	1,12	10,25	0,50	0,37	0,06	0,05	98,03
0,00001	70,14	2,69	0,53	17,13	0,69	8,40	0,45	0,36	0,04	0,20	100,62
0,00001	66,12	2,16	0,53	19,61	0,47	5,77	0,57	0,02	0,15	0,06	95,44
0,00002	71,00	1,89	0,33	20,35	0,43	5,42	0,34	0,01	0,08	0,04	99,90
0,00002	71,40	1,93	0,32	14,82	0,33	5,69	0,38	0,02	0,08	0,04	95,00
0,00001	71,48	1,89	0,30	18,09	0,34	4,74	0,36	0,03	0,04	0,01	97,28
0,00002	66,23	2,02	0,51	14,34	3,64	8,53	1,04	0,30	0,07	0,08	96,76
0,00002	65,73	1,90	0,31	19,24	0,49	5,64	0,45	0,19	0,08	0,01	94,04
0,00002	70,22	1,82	0,25	19,04	0,40	5,08	0,32	0,01	0,08	0,01	97,23
0,00001	71,51	1,89	0,28	18,32	0,41	4,67	0,31	0,01	0,04	0,01	97,45
0,00001	69,86	2,17	0,54	16,87	0,69	7,31	0,57	0,49	0,10	0,13	98,73
0,00001	69,31	2,21	0,53	16,03	0,60	7,82	0,58	0,92	0,09	0,14	98,23
0,00002	68,11	1,87	0,85	17,68	0,90	7,28	0,90	0,29	0,14	0,15	98,17
0,00001	71,77	1,55	0,36	18,38	0,35	5,01	0,42	0,02	0,09	0,06	98,01
0,00002	71,35	1,70	0,42	17,60	0,47	6,06	0,42	0,02	0,10	0,05	98,19
0,00003	66,37	2,41	1,33	16,43	1,02	7,65	2,34	0,59	0,20	0,34	98,68
0,00001	73,48	1,65	0,36	15,71	0,53	5,94	0,34	0,03	0,09	0,15	98,28
0,00001	69,11	2,59	0,51	15,28	0,57	8,03	0,53	1,52	0,09	0,14	98,37
0,00001	69,77	1,93	0,54	17,30	0,64	7,01	0,57	0,35	0,10	0,16	98,37
0,00001	71,30	1,72	0,45	17,22	0,42	6,32	0,55	0,12	0,10	0,12	98,32
0,00001	70,92	1,87	0,62	17,71	0,57	5,97	0,62	0,03	0,11	0,05	98,47
0,00001	69,15	1,70	0,51	19,07	0,43	6,64	0,66	0,03	0,10	0,05	98,34
0,00001	66,09	1,78	1,12	19,33	0,29	7,57	0,81	1,04	0,14	0,13	98,30
0,00002	69,34	2,95	0,74	14,78	0,82	9,09	0,68	0,22	0,10	0,13	98,84
0,00002	67,41	2,58	0,78	15,40	0,73	9,06	0,80	0,37	0,12	0,12	97,37
0,00001	66,84	2,69	0,83	15,33	0,76	9,42	0,91	0,37	0,13	0,12	97,40
0,00001	70,94	2,42	0,48	16,04	0,74	7,12	0,54	0,04	0,10	0,14	98,56
0,00001	65,40	1,39	1,77	17,52	0,39	8,27	0,62	2,98	0,09	nd	98,43
0,00001	68,00	1,75	1,01	19,20	0,54	7,07	0,52	0,03	0,01	nd	98,13
0,00002	65,37	2,34	2,22	18,50	0,50	7,73	0,88	0,48	0,25	nd	97,91
0,00002	70,01	2,15	0,46	17,12	0,61	6,87	0,50	0,54	0,09	0,06	98,41
0,00001	64,30	1,35	2,06	17,55	0,38	8,31	0,57	4,33	0,07	nd	98,92
0,00001	62,38	2,58	1,87	20,41	0,38	6,04	1,12	2,72	0,62	0,08	98,20
0,00001	63,76	3,18	3,77	15,96	0,44	5,63	1,29	1,84	0,59	0,13	96,59
0,00001	63,82	2,22	2,40	18,10	0,75	7,12	1,21	3,14	0,42	0,11	99,29
0,00001	65,35	2,22	0,66	20,34	0,52	6,39	0,71	1,12	0,17	0,01	97,49

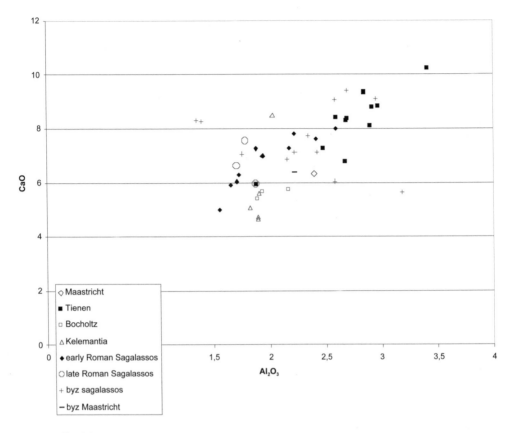

Fig. 2.1
CaO-Al$_2$O$_3$ biplot of the glass analysed in Degryse *et al.* (this volume)

A great deal of the 1st–5th century AD natron glass from the sites studied is distinct from the glass of 4th–8th century primary production centres in the Levant and Egypt. Compositions for comparison with our own analyses were taken from Freestone *et al.* (2000), Nenna *et al.* (2000), Freestone (2006) and Freestone *et al.* (2005). The glass from Sagalassos, Maastricht, Bocholtz and Kelemantia and some of the glass from Tienen dated to the first half of the 5th century AD has in general higher Na$_2$O and lower Al$_2$O$_3$ and CaO contents than the Levantine I, Levantine II and Egyptian II groups. Also, it has lower MgO and SiO$_2$ contents than the Levantine I and II glass and higher MgO and K$_2$O contents than the Egyptian II glass. In comparison to the Egyptian I group, the early and late Roman glass of Sagalassos has higher SiO$_2$, CaO and K$_2$O contents and lower Al$_2$O$_3$ and Na$_2$O contents. The glass from Tienen on the high end of the Al$_2$O$_3$-CaO diagram, however, shows a good correspondence with the Levantine I glass group. The early to late Roman natron glass from the sites studied is, however, very similar in overall composition to early Roman material from all over the empire as defined by Nenna *et al.* (1997), Aerts *et al.* (2000) and Freestone *et al.* (2005).

The possible 1st–3rd century AD plant ash glasses KEL2 and SAG 573 have a major element composition which does not correspond to known plant ash glasses, though there are few data for such Roman 'plant ash' glasses. In view of the elevated but not very high content of K_2O and the high content of Na_2O and Al_2O_3 in this sample, it is probable that this vessel was produced from a mixture of natron and plant ash glass. Similar levels of potassium oxide have been found in early Roman glasses from Fishbourne (Henderson in press). Remarkably, the compostion of KEL2 resembles that of the Egyptian II glass group.

The Sagalassos glass dated to the second half of the fifth up to the seventh century AD corresponds well to that from the known primary production sites of that time. The HIMT glass from Sagalassos is identical to the HIMT group described by Freestone (2006), while most of the other glass from Sagalassos corresponds well to the Levantine I group. The glass from Maastricht, however, does not correspond to the Levantine I group, and has a composition identical to the 1st–3rd century Roman glass. Remarkably, samples SA04VL8A & B correspond well to the Egypt II group (e.g. Freestone 2006).

Raw glass from the 4th–8th century AD primary production sites in Egypt and the Levant has already been analysed for its Sr and Nd isotopic composition (Freestone *et al.* in press). Strontium is considered a proxy for the lime-rich component(s) in the glass raw materials (Freestone *et al.* 2003). In Levantine samples, the $^{87}Sr/^{86}Sr$ signature close to the modern day marine signature of 0.7092 indicated the use of shell as a lime source in the glass (Wedepohl and Baumann 2000, Freestone *et al.* 2003). This shell was a natural inclusion in the beach sand of the Levantine coast, which was used to manufacture the glasses (Brill 1988). The lower $^{87}Sr/^{86}Sr$ signature of the Egyptian samples pointed to either the use of limestone (Freestone *et al.* 2003) or the influence of other minerals in the sand (Degryse *et al.* 2005, Freestone *et al.* 2005) relatively low in radiogenic strontium. The low variation in $^{143}Nd/^{144}Nd$ for Levantine and HIMT (Egyptian) primary glass, with values between $\varepsilon = -6.0$ to -5.1, was consistent with the values given for Nile-dominated sediments in the Eastern Mediterranean (Weldeab *et al.* 2002, Stanley *et al.* 2003). This range of Sr and Nd isotopic values is repeated in most of the 5th–7th century AD glass in this study, confirming its eastern Mediterranean origin. In this way, the main element and isotopic data concur. However, both techniques are complementary and can give different information. Where main elements are less likely to be able to distinguish, for instance, between the use of different sands along the coast of Syro-Palestine (the so-called Levantine I group), the variation in Nd isotopic signature may be more revealing. Conversely, the range in Nd signatures will be similar for the Egyptian or Levantine origin of a glass, while the main element composition of this glass will clearly distinguish Egyptian and Syro-Palestinian sources. The Sr isotopic signature of such glass will indicate the difference in lime source used to make the glass, the low Sr isotope values seem to be indicative of an Egyptian glass origin, while higher values, close to the modern-day oceanic signature, seem to be typical for but not exclusive to a Levantine or Syro-Palestinian origin. In this respect, samples VL8A and VL8B from Sagalassos are remarkable, as their major element composition points to the Egyptian II group, their Nd signature is indicative of an eastern Mediterranean origin, but their Sr signature is close to that of modern-day sea shell, probably identifying the lime source as such.

Some 1st–3rd century AD glass from Sagalassos, Tienen, Bocholtz and Kelemantia has a Sr-Nd isotopic composition identical or very similar to the signature of the known 4th–8th century AD primary production locations in the Levant and Egypt. As mentioned before, the discovery of early Roman glass furnaces in Beirut shows that Early Roman primary glass production took place in the eastern Mediterranean, although not necessarily in the same geographical area as the aforementioned primary glass units, especially for samples with an Nd isotopic signature between –4.4 and -2.5 ε Nd. Such variation in type/composition and geographical location of sands used along e.g. the coast of Syro-Palestine for primary production could therefore be the explanation for the varying major element chemistry between early Roman and later glass, as suggested in previous studies (Nenna *et al.* 1997, Picon and Vichy 2003, Foy *et al.* 2003). The Sr signature of this glass is very homogeneous, between 0.70877 and 0.70905 ^{87}Sr/^{86}Sr. This is nearly identical to the Sr signature of the sands and is due to shell as a lime source of the glass

Conversely, some glass samples from Maastricht, Tienen, Bocholtz and Kelemantia clearly have an exotic Sr-Nd isotopic composition, not corresponding to sediment signatures from the eastern Mediterranean basin. It is clear from the study of e.g. Goldstein *et al.* (1984), Grousset *et al.* (1988) and Weldeab *et al.* (2002) that the Sr and Nd ratios of sediments in the Mediterranean vary significantly. Sediments in the east-west axis of the Mediterranean range from -10.1 ε Nd at Gibraltar to -3.3 ε Nd at the mouth of the river Nile, with a maximum of +4.6 ε Nd of the Graeco–Turkish coast. Samples with an isotopic signature of between -6.4 and -10.8 ε Nd are not consistent with any sediment in the eastern Mediterranean but correspond well to the range in isotopic values of beach and deep-sea sediments from the western Mediterranean, from the Italian peninsula to the French and Spanish coast and from north-western Europe (Degryse and

Schneider 2008). The primary production location of this glass is therefore most likely to lie in the western Roman Empire. The different major element composition of the 1st–3rd century vessel glass as compared to the typical composition of the known 4th–8th century primary producers indicates that the glass has an entirely different primary origin, and is not just a variation in composition of sands in the same geographical area.

Also unusual here is the 2nd century AD glass from Tienen, where a larger part of the glass samples correspond in their major element composition to the 4th–8th century AD Levantine I group. The Nd isotopic signature of these samples does not indicate an eastern Mediterranean origin, but places their primary production location in the western Mediterranean or north-western Europe (Degryse and Schneider 2008). In this example it is difficult to suggest the origins of the glass raw materials using the major element databases currently available.

The Sr signature of most of this glass is very homogeneous, close to the modern-day oceanic composition and likely to be indicative of the use of shell as a lime source in the glass. Some samples show a truly exotic Sr-Nd signature. Sample TIE 49 has a signature of -7.6 ε Nd and 0.70759 $^{87}Sr/^{86}Sr$. This is consistent with the Nd signature of Egyptian sands (Degryse and Schneider 2008) and the earlier analysis of early-Byzantine/Islamic Egyptian glass (Freestone *et al.* in press, Degryse *et al.* 2006a). This could suggest that the glass originated in Egypt. The major element composition of the sample distinguishes it from all other samples and early Roman glass.

Sample KEL 234/88 has a signature of -9.0 ε Nd and 0.70966 $^{87}Sr/^{86}Sr$. The Nd signature of this glass sample suggests an origin in the western Roman Empire (Degryse and Schneider 2008). The Sr signature points to the use of a lime source other than shell or limestone, with an Sr signature relatively higher in radiogenic strontium than the modern seawater composition. The main element composition of this sample is identical to the main early Roman glass group.

The 4th–5th century AD glass from Sagalassos on the one hand has a Sr-Nd isotopic composition identical to the signature of the known 4th–8th century AD primary production locations in the Levant and Egypt. Its main element composition, however, is closer to the early Roman glass group than the Levantine I group. One sample is quite remarkable, with a very exceptional signature of -13.0 ε Nd and 0.7254 $^{87}Sr/^{86}Sr$. Sediments dominated by input from wind-blown Saharan dusts show a typical isotopic composition with ε Nd between -12 to -13.5 and $^{87}Sr/^{86}Sr$ around 0.725 (Goldstein *et al.* 1984). It is tempting to assign the primary origin of this glass on this basis to North Africa, though on the basis of this one analysis this is speculative. The main element composition of this sample is typical of early Roman glass.

Conclusion

Neodymium and strontium isotopes are clearly useful for tracing the origin of primary glass. Nd is characteristic of the mineral fraction other than quartz in the silica raw material, while Sr is in most cases characteristic of the lime component, either attributed to the sand raw material or as a separate constituent in the form of shell. These isotopes do not supplant main element analyses and both techniques discussed here should be regarded as complementary.

In summary: Eastern Mediterranean 4th–8th century AD primary glass has a Nile-dominated Mediterranean Nd signature (higher than -6.0 ε Nd), Syro-Palestinian glass has a sea shell-dominated Sr isotopic signature (close to 0.7092), and (Egyptian) HIMT glass has lower Sr isotopic values (as low as 0.7075). In general, lower $^{87}Sr/^{86}Sr$ values may be indicative of an Egyptian origin for glass (see also Freestone *et al.* 2003). In this period, groups and geographical origins defined by main element analysis (especially Levantine I and HIMT glass) concur well with groups and origins defined on the basis of isotopic data.

Assigning the primary origin of 1st–3rd century AD glass appears not to be as straightforward as for the later period of natron glass production. Some 1st–3rd century AD glass has a Nile-dominated Mediterranean Nd signature (higher than -6.0 ε Nd), pointing to an Eastern Mediterranean origin. This suggests that the glass may have come from primary glassmaking sites in Egypt or in the Levant (Kouwatli *et al.* 2008). Glass with a different Nd signature (lower than -7.0 ε Nd) has also been identified. This signature locates primary production in the western Mediterranean or north-western Europe (Degryse and Schneider 2008).

Moreover, it has also been suggested that some 4th–5th century AD glass may have a North African origin, using Saharan sands. With the current data available, such a mismatch between major element characterisation and the results from Sr and Nd isotopes is difficult to interpret. For example, glass from 2nd century AD Tienen, has a major element composition identical to that of Levantine I glass (but chronologically produced at least two centuries earlier), and has Nd signatures excluding the use of eastern Mediterranean sands. It is unclear, however, how commonly primary glass from outside the eastern Mediterranean was used and on what scale 'western' glass was produced and traded.

Acknowledgements

This research was supported through a Fellowship of the Alexander von Humboldt Foundation awarded to P. Degryse. This research is also supported by the

Interuniversity Attraction Poles Programme - Belgian Science Policy (IUAP VI). The text also presents results of GOA 2007/02 (Onderzoeksfonds K.U.Leuven, Research Fund K.U.Leuven) and of FWO projects no. G.0421.06, G.0585.06 and KAN2006 1.5.004.06N.

References

A. Aerts, B. Velde, K. Janssens, W. Dijkman, 2003, Change in silica sources in Roman and post-Roman glass, Spectrochimica Acta part B, 58, 659-667.

J.L. Banner, 2004, Radiogenic isotopes: systematics and applications to earth surface processes and chemical stratigraphy, Earth Science Reviews, 65, 141-194.

R. H. Brill, 1988, Scientific investigations of the Jalame glass and related finds, in: G.D. Weinberg (ed.) Excavations at Jalame. Site of a glass factory in Late Roman Palestine, Missouri Press, 257-294.

R.H. Brill, 1999, Chemical analyses of early glasses, Corning Museum of Glass.

R.H. Brill, 2006, Strontium isotope analysis of historical glasses and some related materials: a progress report, paper presented at the 17th international conference of the Association Internationale de l'Histoire de Verre, 4-8 September 2006, Antwerp.

R.H. Brill, J.M. Wampler, 1965, Isotope studies of ancient lead, American Journal of Archaeology, 71, 63-77.

W.H. Burke, R.E. Denison, E.A. Hetherington, R.B. Koepnick, H.F. Nelson, J.B. Otto, 1982, Variation of seawater $^{87}Sr/^{86}Sr$ throughout Phanerozoic time, Geology, 10, 516-519.

P. Cosyns, M. Martens, 2002-03, Un four de verrier Romain du deuxième siècle à Tirlemont (Belgique). Bulletin de l'association française pour l'archéologie du verre 2002-03, 34-37.

T. de Groot, 2006, Resultaten van de opgraving van een Romeins tumulusgraf in Bocholtz (Simpelveld), Rapportage Archeologische Monumentenzorg, 127, Rijksdienst voor Oudheidkundig Bodemonderzoek.

P. Degryse, Ph. Muchez, J. Naud, M. Waelkens, M., 2003, Iron production at the Roman to Byzantine city of Sagalassos: an archaeometrical case study, in: Proceedings of the International Conference on Archaeometallurgy in Europe, 24 to 26 September 2003, Milan, Italy, 133-142.

P. Degryse, J. Schneider, J. Poblome, Ph. Muchez, U. Haack, M. Waelkens, 2005, Geochemical study of Roman to Byzantine Glass from Sagalassos, Southwest Turkey, Journal of Archaeological Science, 32, 287-299.

P. Degryse, J. Schneider, U. Haack, V. Lauwers, J. Poblome, M. Waelkens, Ph. Muchez, 2006a, Evidence for glass 'recycling' using Pb and Sr isotopic ratios and Sr-mixing lines: the case of early Byzantine Sagalassos, Journal of Archaeological Science, 33, 494-501.

P. Degryse, J. Schneider, V. Lauwers, 2006b, Sr and Nd isotopic provenance determination of ancient glass, paper presented at the 17th international conference of the Association Internationale de l'Histoire de Verre, 4-8 September 2006, Antwerp.

P. Degryse, J. Schneider, 2008, Pliny the Elder and Sr-Nd isotopes: tracing the provenance of raw materials for Roman glass production, Journal of Archaeological Science, 35, 1993-2000.

P. Degryse, J. Schneider, N. Kellens, M. Waelkens, Ph. Muchez, 2007, Tracing the Resources of Iron Working at Ancient Sagalassos (SW Turkey): a Combined Lead and Strontium Isotope Study on Iron Artefacts and Ores, Archaeometry, 49, 75-86.

W. Dijkman, 1993, La terre sigillée décorée à la molette à motifs chrétiens dans la stratigraphie maastrichtoise (Pays-Bas) et dans le nord-ouest de l'Europe, in: Gallia, 49, 129-172.

G. Faure, 1986, Principles of isotope geology, 2nd ed, John Wiley and Sons.

G. Faure, 2001, Origin of igneous rocks, the isotopic evidence, Springer.

D. Foy, M. Vichy, M. Picon, 2000, Lingots de verre en Méditerrané occidentale, in : Annales du 14th congrès de l'Association pour l'Histoire du Verre, AIHV, Amsterdam, 51-57.

D. Foy, M. Picon, M. Vichy, V. Thirion-Merle, 2003, Charactérisation des verres de la fin de l'Antiquité en Mediterranée occidentale: l'émergence de nouveaux courants commerciaux, in : D. Foy, M.D. Nenna (eds.) Echanges et Commerce du verre dans le Monde Antique, Editions Monique Mergoil, 41-85.

I. C. Freestone, Y. Gorin-Rosen, M. J. Hughes, 2000, Primary glass from Israel and the production of glass in Late Antiquity and the Early Islamic period, in: M.D. Nenna (ed.) La route du verre. Ateliers primaires et secondaires du second millénaire avant J.C. au Moyen Age, Travaux de la Maison de l'Orient Méditerranéen, 33, TMO, 65-82.

I.C. Freestone, R. Greenwood, M. Thirlwall, 2002a, Byzantine and early Islamic glassmaking in the eastern Mediterranean: production and distribution of primary glass, in: Kordas, G. (ed.), Proceedings of the 1st international conference on Hyalos-Vitrum-Glass. History, technology and restoration of glass in the Hellenic world, 167-174.

I.C. Freestone, M. Ponting, J. Hughes, 2002b, The origins of Byzantine glass from Maroni Petrera, Cyprus, Archaeometry 44, 257-272.

I.C. Freestone, K. A. Leslie, M. Thirlwall, Y. Gorin-Rosen, 2003, Strontium isotopes in the investigation of early glass production: Byzantine and early Islamic glass from the Near East, Archaeometry, 45, 19-32.

I.C. Freestone, S. Wolf, M. Thirlwall, 2005, The production of HIMT glass: elemental and isotopic evidence, in: Proceedings of the 16th Congress of the Association Internationale pour l'Histoire du Verre, London, 153-157.

I.C. Freestone, 2006, Glass production in Late Antiquity and the Early Islamic period: a geochemical perspective, in: M. Maggetti, B. Messiga (eds) Geomaterials in Cultural Heritage, Geological Society of London Special Publication, 201-216.

I.C. Freestone, P. Degryse, J. Shepherd, Y. Gorin-Rosen, J. Schneider, in press, Neodymium and Strontium Isotopes Indicate a Near Eastern Origin for Late Roman Glass in London, Antiquity.

S.L. Goldstein, R.K. O'Nions, P.J. Hamilton, 1984, A Sm-Nd isotopic study of atmospheric dusts and particulates from major river systems, Earth and Planetary Science Letters, 70, 221-236.

J. Henderson, 1989, The scientific analysis of ancient glass and its archaeological interpretation, in: J. Henderson (ed) Scientific Analysis in Archaeology, Oxford University Committee for Archaeology, 30–62.

J. Henderson, 1993, Aspects of early medieval glass production in Britain, in: Proceedings of the 12ᵗʰ Congress of the International Association of the History of Glass, 26-31.

J. Henderson, in press, The provenance of archaeological plant ash glasses, in: A.J. Shortland, Th. Rehren, I.C. Freestone (eds) From mines to microscope - Studies in honour of Mike Tite, University College London Press.

J. Henderson, D. McPhail, 2004, Radical changes in Islamic glass technology: evidence for conservatism and experimentation with new glass recipes from early and middle Islamic Raqqa, Syria, Archaeometry, 46, 439-468.

J. Henderson, J.A. Evans, H.J. Sloane, M.J. Leng, C. Doherty, 2005, The use of oxygen, strontium and lead isotopes to provenance ancient glasses in the Middle East, Journal of Archaeological Science, 32, 665-673.

J. Henderson, K. Challis, S. O'Hara, S. McLoughlin, A.Gardner, G. Priestnall, 2005a, Experiment and innovation: early Islamic industry at al-Raqqa, Syria, Antiquity, 79, 130–145.

J. Henderson, J. Evans, Y.Barkoudah, in press, The roots of provenance: glass, plants and isotopes in the Islamic Middle East, Antiquity.

C. Pin, D. Briot, C. Bassin, F. Poitrasson, 1994, Concomitant separation of strontium and samarium-neodymium for isotopic analysis in silicate samples, based on specific extraction chromatography, Analytica Chimica Acta, 298, 209-217.

D.J. Huisman, T. de Groot, S. Pols, B.J. van Os, P. Degryse, in press, Compositional variation in Roman colourless glass objects from the Bocholtz burial (The Netherlands), Archaeometry.

C. Isings, 1957, Roman glass from dated finds, Archaeologica Traiectina II.

C. Jackson, 1997, From Roman to early medieval glasses. Many happy returns or a new birth?, in: Proceedings of the 13ᵗʰ Congress of the International Association for the History of Glass, 289-302.

R. Jackson, R. Burchill, D. Dungworth, C. Mortimer, 2005, Excavations on the site of Sir Abraham Elton's glassworks, Cheese Lane, Bristol, Post-Medieval Archaeology, 39, 92-132.

C.M. Jackson, L. Joyner, C.A. Booth, P.M. Day, E.C.W. Wager, V. Kilikoglou, 2003, Roman Glass-Making At Coppergate York? Analytical Evidence For The Nature Of Production, Archaeometry, 45, 435-456.

I. Kouwatli, H.H. Curvers, B. Sturt, Y. Sablerolles, J. Henderson, P. Reynolds, 2008, A pottery and glass production site in Beirut (015), BAAL, 10.

K.A. Leslie, I.C. Freestone, D. Lowry, M. Thirlwall, 2006, Provenance and technology of near Eastern glass: oxygen isotopes by laser fluorination as a compliment to Sr, Archaeometry, 48, 253-270.

M.D. Nenna, M. Vichy, M. Picon, 1997, L'Atelier de verrier de Lyon, du Ier siècle après J.-C., et l'origine des verres "Romains", Revue d'Archèomètrie, 21, 81-87.

M.D. Nenna, M. Picon, M. Vichy, 2000, Ateliers primaires et secondaires en Égypte à l'époque gréco-romaine, in: M.D. Nenna (ed.) La route du verre. Ateliers Primaires et Secondaires du Second Millénaire av. J.-C. au Moyen Âge, Travaux de la Maison de l'Orient Méditerranéen, 33, 97-112.

M.D. Nenna, 2003, Les ateliers égyptiens à l'époque gréco-romaine, in: D. Foy (ed.) Coeur de verre. Production et diffusion du verre antique, Gollion, 32-33.

P.T. Nicholson, C.M. Jackson, K.M. Trott, 1997, The Ulu Burun glass ingots, cylindrical vessels and Egyptian glass, Journal of Egyptian Archaeology, 83, 143-153.

T. Panhuysen, W. Dijkman, 1987, Romeinse vondsten van de Scharnweg, in: Publications de la Société Archéologique et Historique dans le Limbourg, CXXIII, 212-216.

M. Picon, M. Vichy, 2003, D'Orient en Occident: l'origine du verre à l'époque romaine et durant le haut Moyen Âge, in: D. Foy, M.D. Nenna (eds.) Echanges et commerce du verre dans le monde antique, Monographies Instrumentum, 24, éditions Monique Mergoil, 17-31.

M. Pollard, in press, What a long strange trip it's been: lead isotopes and Archaeology, in: A.J. Shortland, Th. Rehren, I.C. Freestone (eds.) From mines to microscope - Studies in honour of Mike Tite, University College London Press.

Th. Rehren, E.B. Pusch, 1997, New Kingdom glass- melting crucibles from Qantir-Piramesses, Journal of Egyptian Archaeology, 83, 127-141.

E.V. Sayre, R.V. Smith, 1961, Compositional categories of ancient glass, Science, 133, 1824-1826.

A. J. Shortland, 2004, Evaporites of the Wadi Natrun: seasonal and annual variation and its implication for ancient exploitation, Archaeometry, 46, 497-516.

A.J. Shortland, M.S. Tite, I. Ewart, 2006, Ancient exploitation and use of cobalt alums from western oases of Egypt, Archaeometry, 48, 153-168.

A. Silvestri, G. Molin, G. Salviulo, R. Schievenin, 2006, Sand for Roman glass production: an experimental and philological study on source of supply, Archaeometry, 48, 415-432.

J.D. Stanley, M.D. Krom, R.A. Cliff, J.C. Woodward, 2003, Nile Flow Failure at the end of the Old Kingdom, Egypt: Strontium Isotopic evidence, Geoarchaeology, 18, 395-402.

M. Walton, 2005. A materials chemistry investigation of archaeological lead glazes. Doctorate Thesis, University of Oxford, Linacre College.

K.H. Wedepohl, A. Baumann, 2000, The use of marine molluscan shells in the Roman glass and local raw glass production in the Eifel area (Western Germany), Naturwissenschaften, 87, 129-132.

K.H. Wedepohl, W. Gaitzsch, A.B. Follmann-Schulz, 2003, Glassmaking and glassworking in six Roman factories in the Hambach Forest, Germany, in: Annales of the 15th Congress of the Association Internationale de l'Histoire de Verre, New York, 53-55.

S. Weldeab, K.C. Emeis, C. Hemleben, W. Siebel, 2002, Provenance of lithogenic surface sediments and pathways of riverine suspended matter in the eastern Mediterranean Sea: evidence from $^{143}Nd/^{144}Nd$ and $^{87}Sr/^{86}Sr$ ratios, Chemical Geology, 186, 139-149.

S. Wolf, S. Stos, R. Mason, M.S. Tite, 2003, Lead isotope analyses of Islamic pottery glazes from Fustat, Egypt, Archaeometry, 45, 405-420.

The provenance of Syrian plant ash glass: an isotopic approach

Julian Henderson, Jane Evans, Youssef Barkoudah

Introduction

The Islamic world during the 'Abbasid caliphate in the 8th and 9th centuries AD can be regarded as one of the most highly centralized periods of ancient Islamic civilisation. It was during this period, and partly also during the preceding Ummayad caliphate, that Islamic material culture emerged as recognisably diagnostic and distinct from neighbouring Byzantine material culture. Technological innovation occurred alongside advances in scholarship, arts and sciences. It was under Harūn al-Rashid and his son al-Mamun that the 'translation movement' occurred. Latin and Greek scientific and other texts were translated, and associated experimental work was pursued. During the 'Abbasid caliphate the Islamic domain stretched from southern Spain to central Asia. The trading network linked the Mediterranean Sea to the Indian Ocean and as far as China via the Silk Road. Craftsmen travelled across this enormous area and goods were traded along it. The *suqs* (markets) in urban centres provided local nodes for the distribution of manufactured metal, pottery and glass goods, sometimes being the end points for traded glass vessels and glass raw materials.

Evidence of the trade in glass is provided by the excavation of the early 11th century AD Serçe Limani shipwreck off the coast of Turkey. Here two metric tonnes of raw glass blocks of up to 30 cm across (Bass 1984, 65) and 11,000 glass lid moils (Bass 1984, 67, Fig. 4) were found, all of which could have derived from the Levantine coast. Such glass may have been used in urban centres such as Istanbul for remelting, perhaps in the production of (Byzantine) glass mosaic tesserae (Andreescu-Treadgold and Henderson 2007). Moreover, glass raw materials were traded. In 985 AD the Arabic geographer, al-Mukaddasi, mentions that ūshnan (*kali*) - plant ash - was an export from the province of Aleppo for glass production (Ashtor 1992, 482). The manufacture of such glass occurred in primary glassmaking centres.

The natron glass which was manufactured in the 7th century AD, presumably under Byzantine auspices, would have been used for the manufacture of glass vessels used in Islamic (Byzantine and Jewish) social contexts. It was fused in large tank furnaces at Levantine coastal sites like Apollonia, Hadera and Dor (Gorin-Rosen 2000). After *c*. 800 AD a 'new' kind of glass, made from a mixture of halophytic plant ashes and silica, was introduced (Henderson 2002). The introduction of this 'new' glass formed part of the emergence of a strong Islamic cultural identity and would have involved new technological procedures, or at the very least experimentation with the working properties of the glass produced. Historical evidence indicates that the Qadisiyya quarter of Baghdad was a production centre (Lamm 1929–1930, 498). There are also historical references to the manufacture of glass in Tyre (Carboni *et al*. 2003). Moreover, that al-Raqqa was an important cosmopolitan 'Abbasid industrial centre where both Christians and Muslims lived is supported historically (Heidemann 2003, 2006). Using variations in vessel form and decoration Lamm (1929–1930) suggested that Ayyūbid and Mamluk *enamelled* vessels were made in three Syrian cities: Damascus, Aleppo and al-Raqqa, but there is no archaeological evidence that supports this. One way of investigating Lamm's suggestion is by scientifically analysing the glass.

The investigation of Islamic glass production using scientific techniques aims to provide answers to a range of key research questions. Amongst these are (1) How widely distributed were the primary glassmaking centres? (2) To what extent was Islamic glass traded from primary glassmaking centres? (3) What was the degree of specialisation in glass production (colouring, vessel forms and decoration)? (4) What was the influence of Byzantine glass production on Islamic glass production? (5) What part, if any, did other technologies (e.g. glazed ceramics) play in the development of glass technology and *vice versa*? (6) Is it possible to produce a geological provenance for Islamic glass? (7) What raw materials were used to make Islamic glass? The answer to question (6), in particular, can help to answer most of the other questions. However, hitherto, although the determination of glass chemical analyses has provided evidence for the raw materials used to make Islamic glass and, where primary glassmaking sites have been found, it has been possible to determine the compositional types of raw glass manufactured there, it has not been possible to provide a geological provenance for the glass. In this article we discuss the isotopic results for Islamic glass from a production site and for plants sampled from Syria and Lebanon. A fuller assessment of technique's potential is published elsewhere (Henderson *et al*., in press).

Glass production at al-Raqqa

The key to proving the existence of Islamic primary plant ash glass production is archaeological excavation. It is important to distinguish between primary production, where glass was fused from primary raw materials, and secondary production, where raw glass was reheated and worked by cold-working, moulding and blowing the glass into vessels and objects. For primary production it is likely that local plants and silica sources were used; a means of tracing the use of local materials in Raqqa products is by determining the isotopic signatures in both raw materials and glass. Archaeological excavations have revealed the evidence of primary plant ash glass production in al-Raqqa, Syria (Henderson 1999, Henderson *et al.* 2005a) and Tyre, Lebanon (Aldsworth *et al.* 2002), and secondary glass production in Beirut, Lebanon (Foy 1996, 2000). Tyre has produced the most complete evidence of tank furnaces used for plant ash glass production. These date to c. 10th–12th centuries AD and had maximum dimensions of 6.4 x 3.9 and 4.5m x 2.2m (Aldsworth *et al.* 2002). It has been estimated that the largest furnace produced a minimum of 37 tons of raw plant ash glass.

Thus far al-Raqqa is the only scientifically excavated Islamic (*inland*) site which has produced evidence for primary *and* secondary glass production. The furnaces were located in an industrial complex of 2 km in length, where glass manufacture occurred alongside glazed and unglazed pottery production (Henderson *et al.* 2005a). Some ten metric tons of tank furnace fragments with raw green, brown, colourless and blue glass attached were excavated from three sites within the complex dating to the 8th–9th, 11th and 12th centuries AD, so primary manufacture occurred intermittently over a period of *c.* 400 years and is supported by the evidence of frit. Secondary production (glass blowing) was carried out in so-called 'beehive' shaped furnaces, the remains of which were found in two workshops, one with the remains of four furnaces and the other with only one. Raw plant ash or natron glass would have been blown into vessels.

Various soda-rich halophytic plants belonging to the *Chenopodiaceae* family could have been fused with silica to make glass in al-Raqqa (and other parts of the Middle East). The plants range from coastal species such as *Salsola soda* to types which grow in semi-desert environments, such as *Salsola kali, S. jordanicola, S. vermiculata, S. Sueda, Anabasis syriaca, Arthrocnemum strobilaceum* and *Haloxylon salicornicum* (Barkoudah and Henderson 2006, Tite *et al.* 2006). The sodium carbonate (alkali) in these plants acted as a flux in glassmaking, reducing the melting temperature of silica from *c.* 1710°C to *c.* 1150°C. The ashed plants would also have introduced some, if not all, of the calcium oxide providing the third major component of the glass (Barkoudah and Henderson 2006). Possible sources of silica would have been crushed quartz, or sand derived from rivers and quarries. When fusing plant ash and silica in a tank furnace the furnace atmosphere (amongst other factors) would have played an important role in determining the final glass colour (Henderson 2000, 30).

The scientific analyses of ancient glasses using techniques such as electron probe microanalysis and inductively-coupled plasma emission spectrometry allow us to define compositional types of glass, and in some cases they can be associated with specific production centres, especially in the Middle East. For example, in Fig. 3.1 we show the range of compositional types found amongst al-Raqqa glasses consisting of two principal plant ash glasses (types 1 and 4), a natron glass (type 3) and a third smaller group of plant ash glass (type 2) (Henderson et al. 2004); the plant ash glasses were all made in situ. The natron glass was imported in its raw form from production centres on the Levantine coast, where it was melted from sand and a mineral alkali, natron (Freestone et al. 2000). The chemical results from 9th century al-Raqqa have provided the widest compositional range from any ancient production site, and their discovery has led to the suggestion that experimentation with raw material recipes occurred there when the technological transition from natron to plant ash glass was occurring (Henderson 2002, Henderson et al. 2004)

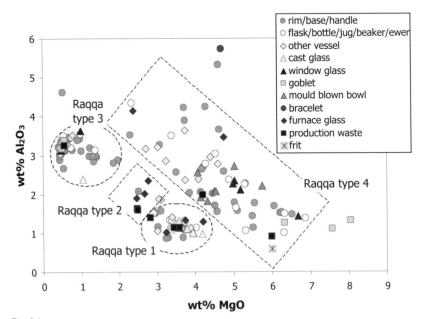

Fig. 3.1
Biplot of MgO-Al$_2$O$_3$, showing the different compositional types of al-Raqqa glass

The alumina levels in Fig. 3.1 partly reflect the impurity variations in silica sources used, higher levels reflecting the presence of alumina-rich minerals in sand and lower levels the use of purer silica sources, such as vein quartz or quartz-bearing sands. Variations in the magnesia levels are a reflection of the alkali (plant ash or natron) sources used, natron glass containing less than 1,5 %. However, at the moment, there is no relationship between chemical compositions and production location that provides a clear *geological/geographical* means of provenancing plant ash glasses. Promising clusters of compositional data for Islamic plant ash glasses are shown in Fig. 3.2. The results plotted are for glass from 9th century AD al-Raqqa (raw glass), facet-cut vessels and goblets from 9th century AD al-Raqqa (Henderson *et al*. 2004); 10th century AD Nishapur (vessels),10th century AD Siraf (vessels) and 9th–10th century AD colourless cameo decorated vessels, 9th century scratch-decorated vessels, 9th century lustre-painted vessels, (Brill 1999). These clusters, based on relative levels of calcium oxide and alumina, suggest that compositionally somewhat different calcium-bearing raw materials were used to make plant ash glass in Syria, the Persian Gulf and Iran; the cluster of data for cameo decorated vessels presumably suggest that we are seeing the result of a single melt. However, it is difficult to use such clusters for the purpose of provenancing glass.

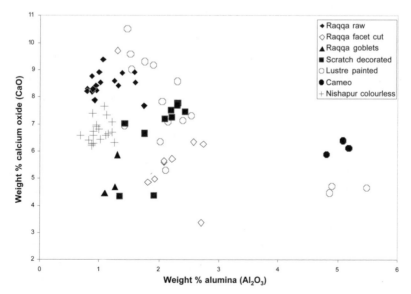

Fig. 3.2
Compositional data for several Islamic plant ash glass samples from 9th century AD al-Raqqa (raw glass and facet-cut vessels and goblets, Henderson *et al.* 2004), 10th century AD colourless Nishapur (vessels), 9th century scratch-decorated and lustre-painted vessels,10th century AD Siraf (vessels) and 9th–10th century AD cameo decorated vessels (Brill 1999)

The evidence of glass production at al-Raqqa therefore provides an ideal means of testing the effectiveness of isotope analysis in the provenance of plant ash glasses. If local raw materials were used to make plant ash glass, and they can be characterized using scientific techniques as having derived from local sources, we can provide a true *provenance* for the glasses and at the same time investigate whether fractionation has occured. By analysing raw furnace glasses we are able to *minimize* the possibility that we are dealing with glasses made in different places and at the same time demonstrate when glass has been imported. Over and above the chemical compositions, isotopic data therefore provide entirely new information about the age and type of raw materials used, with the potential to provide a provenance for glass. The overall aim of the study is to contribute, in fundamental ways, to our understanding of the source of glass raw materials, forming essential components of the *chaîne opératoire* for glass production and leading, eventually, to glass provenance.

The principles of isotope analysis and how isotopes contribute

The determination of strontium (Sr), neodymium (Nd), lead (Pb) and oxygen (O) isotopes in ancient glasses is a relatively recent application, and it has started to show that there is great potential for shedding new light on ancient glass technologies. (Wedepohl and Baumann 2000, Freestone *et al.* 2003, Henderson *et al.* 2005b). Here we discuss the first set of results for both Sr and Nd isotope determinations of Islamic plant ash glasses. Furthermore, we present the first set of Sr data which establishes variations in bioavailable Sr isotope signatures across northern Syria, in parts of southern Syria and in the Lebanon. By establishing the Sr isotope variation in the landscape it becomes possible to link the geological age of the bedrock in which the Sr is found, and therefore the source of the plant ash, to the glass made from it. It is hoped that such results will eventually lead to the construction of trade patterns of raw glass and diagnostic vessel forms. At the very least it will prove when a glass sample was *not* made in Syria/Lebanon.

The $^{87}Sr/^{86}Sr$ ratio depends upon the age and rubidium content of the rock. Since strontium generally substitutes for calcium, Sr isotopes will provide data on the source, nature and age of lime. For example, lime from old marine limestones can be distinguished from modern-day seashell sources (in sand). Alkali-rich plants take up Sr from the soil in which they grow and the $^{87}Sr/^{86}Sr$ ratio is a reflection of the age of the underlying geology (Freestone *et al.* 2003). This means that the Sr isotope composition of plant ash glasses provides information about the geological/surfical environment in which the plant grew. When the plant is ashed and used for the manufacture of glass, the $^{87}Sr/^{86}Sr$ is passed on to the glass without fractionation (Evans and Tatham 2004, Sillen *et al.* 1998), and this is shown in this chapter by the Sr results obtained. Assuming that the plants used for glass production grew close to the glassmaking site, the Sr isotope signature is carried from the soil, via the plants to the glass, and this becomes a way of providing a geological provenance for the glass made at the production site (Henderson *et al.* 2005b). The *concentration* of Sr in the plants can vary according to a range of factors, of which the geology is a significant one (Barkoudah and Henderson 2006). For the technique to be successful as a provenancing tool it relies on a minimum amount of glass mixing – especially with plant ash glasses made in other areas with different Sr isotope signatures. Also, in the unlikely event that plants gathered by Bedouins from geological environments with contrasting ages were mixed, it would be possible to detect this isotopically when compared to Sr results for plants sampled from known locations. If scrap glass was added to the melt at secondary production centres where vessels were blown, even here it is possible to demonstrate using isotopic analysis that mixing has occurred. As long as there are sufficient samples of 'unmixed' raw glass it is possible to identify mixing, because one isotope signature will 'dilute' the other (Henderson *et al.* 2005b, Degryse *et al.* 2006). The same isotopic signatures will be obtained irrespective of the species or preparation technique as long as they are growing in an area dominated by the same bedrock geology. Moreover, performing Sr isotope determinations on plants growing at various locations in northern Syria has established Sr isotope variation, so that we are also able to assess the potential of detecting other glass production centres.

Nd is geologically linked to silica and silicate minerals that were used to make glass. Although silica (quartz) in a pure form contains negligible rare earth elements, impurities within vein quartz, or other minerals such as feldspar within a sand source, will provide a mineralogical trace for the silica source. Moreover, the ratio of the radiogenic isotope $^{143}Nd/^{144}Nd$ is a reflection of the formation age of the silica and the concentration of parent element, samarium. Therefore contrasting ages of geological sand deposits can be detected in glass made from them. It should therefore be possible to discriminate between a sand derived from the erosion of old, metamorphic rocks and one derived from the erosion of a younger rock, such as a granite. The concentration of Nd within glass can also be used as an indication of silica source type, since the purer the silica source the lower the impurities and, hence, the Nd concentration. The Nd levels detected in halophytic plants have been found to be negligible (Barkoudah and Henderson 2006, Table 2) so they are unlikely to contribute significantly to the isotopic signature.

Clayton and Mayeda (1963a, b) suggested that the determination of oxygen isotope ratios in silica could provide a means of sourcing the silica, because geology of different ages would produce different $\delta^{18}O$ values. Soda-lime-silica glass, the compositional type that predominates in large parts of the ancient world until c. 1000 AD, contains between c. 60% and 70% silica. Therefore the $\delta^{18}O$ values are likely to be the result of a dominant silica contribution. The other major oxide in these glasses is soda, which is likely to have a somewhat different $\delta^{18}O$ value, and this should be taken into account in interpreting the results of oxygen isotope analysis. Brill (1970) was one of the first to apply this technique to the investigation of ancient glass. Methodological alternatives and improvements have been reported recently (Alexandre *et al.* 2006).

Given the contrast in geologies involved in the areas studied (Beydoun 1977) one would hope that their respective Sr isotope signatures would distinguish between glasses made in each area. Therefore by combining the results of Sr and Nd isotope determinations it is potentially possible to characterize and provenance glasses in ways that may cut across the results of chemical analyses (Henderson in press). Glass chemical compositions provide evidence of the recipes used and impurity levels. Isotopic measurements provide evidence of the geological age of the silica or the calcium component in soil derived from the bedrock (in plant ashes) used.

Methodology

Glass samples were broken into fragments and cleaned by placing in 2.5HCl in an ultrasonic bath for five minutes, followed by three rinses in de-ionized water for the same length of time. The fragments, once dry, were milled to a fine powder in an agate micro ball mill. Standard silicate dissolution was then undertaken and the Sr and REE fractions collected from Dowex resin columns, and Nd was separated from the REE fraction using HDEHP columns (see Dickin 1995 for details). Plant samples were cryogenically milled to produce a powder and were dissolved using Teflon distilled 16M HNO_3 followed by Aristar hydrogen peroxide, then converted to chloride using quartz distilled 6M HCl. Sr was separated using Dowex columns.

Nd and Sr concentrations were determined by isotope dilution using a mixed [150]Nd tracer and enriched [84]Sr solution respectively. Sm and Nd concentrations and [143]Nd/[144]Nd isotope ratios were measured using a double filament assemblage on either a Finnegan Mat 262 or a Thermo Triton multi-collector and Sr was measured using a rhenium single filament with TaF activator after the method of

Birck (1986). The reproducibility of the data is based on standards data run during sample analysis using NBS987 international Sr standard and an 'in-house' J&M standard. Typical results during this period are: Finnigan MAT 262 gave NBS987 = 0.710202 ± 0.000024 (2σ, n=7) and J&M = 0.511193 ± 0.000034 (2σ, n=12); and the Thermo Triton gave NBS987 = 0.710263 ±0.000008 (2σ, n=50) and J&M = 0.511104 ± 0.000006 (2σ, n=32). Data from both machines were normalized to NBS987 = 0.710250 and J&M = 0.511123 which is the calibrated equivalent of La Jolla = 0.511864. All the $^{143}Nd/^{144}Nd$ ratios were determined to ≤ 0.000015 (2SE). Data are corrected to a $^{146}Nd/^{144}Nd$ ratio of 0.7219.

Oxygen isotope ratios were determined using a method similar to that of Clayton and Mayeda (1963a, b). The samples were out-gassed at 250°C, and brought to full reaction with 200mb of ClF_3 at 450°C for 16 hours. Samples were pre-fluorinated before the reaction. The oxygen yields were converted to CO_2 by reaction with platinized graphite rod heated to 675°C by induction furnace. The resultant CO_2 yields were measured with a capacitance manometer. All samples were normalized through NBS 28. Normal experimental error for pure quartz is quoted at ±0.2‰; errors on repeated analyses of some of these glasses were better than 0.2‰. The data are presented on the VSMOW scale following the Vienna calibration approach.

Results

The results for 31 plant samples and 24 glass samples are given in Tables 3.1 and 3.2. Most samples are of 9th century AD date and derive from Tell Zujaj, the glass factory site. Exceptions are two 9th century AD window samples from the west palace complex north of the industrial zone (RAQ 66 and 67) (see Table 3.2); two 11th century AD armlets (RAQ 268 and 269) from Tell Fukhkhar in the industrial complex; and a single 12th century AD vessel fragment (RAQ 61) from the princess's palace within the walls of al-Rafica, the twin city to al-Raqqa located 1 km to its west.

Sample	Location	Sr (ppm)	$^{87}Sr/^{86}Sr$
SYR 01	Bab el Hawa	251	0.708514
SYR 02	Bab el Hawa	271	0.708477
SYR 03	Bab el Hawa	123	0.708503
SYR 04	Aleppo Citadel	608	0.708091
SYR 05	Aleppo Citadel	776	0.708064
SYR 06	Aleppo Park	1070	0.708160
SYR 07	Aleppo Park	87	0.707872
SYR 10	Jaboul salina	515	0.708085
SYR 11	Heracla	55	0.708090
SYR 12	Heracla	256	0.708022
SYR 13	Al-Raqqa	844	0.708254
SYR 14	Al-Raqqa	587	0.708082
SYR 15	Balikh valley	145	0.708402
SYR 16	Balikh valley	33	0.708339
SYR 17	Balikh valley	32	0.708597
SYR 18	Balikh valley	102	0.708265
SYR 19	Tel Brak	71	0.708232
SYR 20	Tel Brak	77	0.708154
SYR 21	Ugarit	141	0.708247
SYR 22	Ugarit	174	0.708396
SYR 23	Wadi Barada (NW Damascus)	45	0.708057
SYR 24	Wadi Barada	121	0.707808
SYR 25	Wadi Barada	51	0.707937
SYR 26	Najha S Damascus	365	0.706890
SYR 27	Najha S. Damascus	338	0.707180
SYR 28	Bab Sharki, Damascus	120	0.707846
SYR 29	Amara SE of Damascus	33	0.707826
SYR 30	NE Damascus	29	0.707815
SYR-8	Jaboul salina	36	0.708020
SYR-9	Jaboul salina	71	0.708069
HEND 19 (L-1)	Lebanon	32	0.707928

Table 3.1
Analytical data for plant ashes analysed (Henderson *et al.* in press)

Sample	Sr ppm	$^{87}Sr^{86}Sr$	Nd ppm	$^{143}Nd/^{144}Nd$	$\delta^{18}O$ V-SMOW	Artefact, date	Compn*
RAQ 33-34c	517	0.708178	5.6	0.511987	13.7	Raw, 9th cent	1
RAQ 33 34M	503	0.708090	5.8	0.512056	14.9	Raw, 9th cent	1
RAQ 45-46	542	0.708148	6.3	0.512102	13.7	Raw, 9th cent	1
RAQ 50	325	0.708208	7.7	0.512204	13.5	Vessel, 9th cent	2
RAQ 54	628	0.708121	6.7	0.512091	13.7	Vessel, 9th cent	1
RAQ 62	494	0.708374	6.6	0.512071	13.1	Vessel, 9th cent	1
RAQQA-26	582	0.708204	5.9	0.512118	13.9	Raw, 9th cent	1
RAQQA-268	469	0.708293	4.1	0.512190	13.7	Bangle, 11th cent	4 mid Al
RAQQA-269	208	0.708289	5.6	0.512203	13.7	Bangle, 11thcent	4 mid Al
RAQQA-35	602	0.708131	5.7	0.512084	13.5	Raw, 9th cent	1
RAQQA-38	361	0.708150	4.0	0.512114	16.6	Raw, 9th cent	4 hi Al
RAQQA-40	515	0.707595	11.0	0.512340	14.9	Raw, 9th cent	2
RAQQA-41	463	0.708448	3.3	0.512148	13.5	Vessel, 9th cent	4 mid Al
RAQQA-42	1026	0.708354	4.1	0.512123	13.9	Raw, 9th cent	1
RAQQA-47	226	0.708175	4.4	0.512107	14	Vessel, 9th cent	4 lo Al
RAQQA-49	726	0.708110	7.1	0.512023	14	Raw, 9th cent	1
RAQQA-58	309	0.708152	2.1	0.512206	13.7	Vessel, 9th cent	4 lo Al
RAQQA-60	648	0.708620	5.9	0.512027	13.2	Raw, 9th cent	1
RAQQA-61	170	0.709347	9.4	0.512084	15	Vessel, 12th cent	2
RAQQA-66	112	0.708556	5.8	0.512244	13.7	Window, 9th cent	4 mid Al
RAQQA-67	426	0.708520	6.3	0.512221	13	Window, 9th cent	4 mid Al
RAQQA-80	70	0.708428	7.4	0.512095	13.5	Vessel, 9th cent	2
RAQQA-81	291	0.708220	3.6	0.512203	13.5	Vessel, 9th cent	1
RAQQA-84	172	0.708150	5.9	0.512035	13.4	Raw, 9th cent	4 mid Al
RAQQA-87	663	0.708129	5.8	0.512118	13.6	Raw, 9th cent	4 hi Al

Table 3.2
Analytical data for the plant ash glasses analysed (Henderson *et al.* in press). Compn* = compositional type as determined by electron probe microanalysis, Fig. 3.1 (see also Henderson *et al.* 2004)

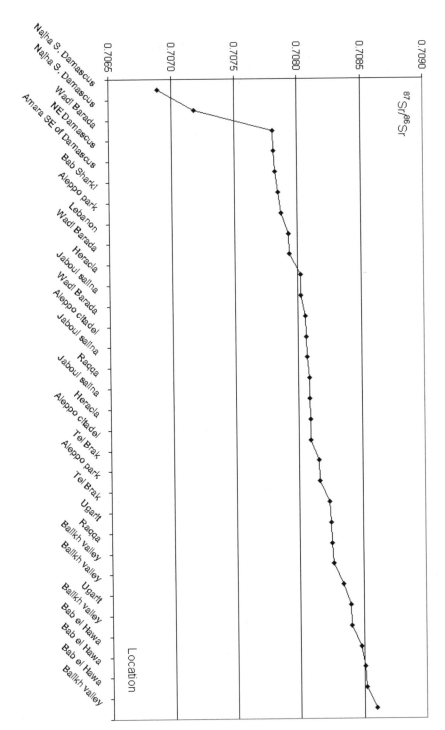

Fig. 3.3
87Sr/86Sr data of plant ashes from several locations in Syria

STRONTIUM

A survey of plant $^{87}Sr/^{86}Sr$ composition was undertaken to attempt to assess the $^{87}Sr/^{86}Sr$ isotope composition that plants from different parts of Syria would contribute to the glassmaking process, and to look for regional variations. Fig. 3.3 shows the $^{87}Sr/^{86}Sr$ values obtained from different sites. The total range of values is not large (0.70689-0.70860) compared with a range currently recorded from British minerals waters of 0.7059-0.7207 which, although not direct biosphere measurements, provide a guide to biosphere variation across the UK (Montgomery *et al.* 2006). The geology of Syria is dominated by Cretaceous, Eocene and Neogene sedimentary deposits. The highest values ($^{87}Sr/^{86}Sr > 0.7084$) in this study are from Eocene and Neogene geology around Bab el Hawa, west of Allepo and in the Balikh valley, close to al-Raqqa and to the north of it (Fig 3.3). Cretaceous rocks would produce a minimum Cretaceous $^{87}Sr/^{86}Sr$ value of 0.7072 (MacArthur *et al.* 2001). However, south-east of Damascus there is a large area of basalt (Asch 2005), and this is probably what is influencing $^{87}Sr/^{86}Sr$ plant composition in this area. The lowest $^{87}Sr/^{86}Sr$ values are from this area: SYR 26 & 27 with 0.7069 and 0.7072 respectively. The restricted range recorded in this study has both advantages and disadvantages; it means that it may not be easy to make clear distinctions between some Sr signatures for glasses made within Syria, but on the wider international scale it should provide a quite well-constrained Syrian signature.

The plant samples were also measured for Sr concentration. This is rather an arbitrary measurement since it depends upon how desiccated the plants were, but it provides a starting point for assessing the Sr contribution of plants to the glassmaking process. Ashing plants will hugely increase the Sr concentration in the ash over the unburned plant material. Plant physiology, the bedrock and drift geology on which the plants grew, and the species and genera of the plants involved can contribute to the concentrations of Sr and other elements in plants (Barkoudah and Henderson 2006). In Fig. 3.4 the glasses used to make a range of artefact types from al-Raqqa (raw furnace glass, vessel glass, window glass and bangles) are plotted using Sr concentration versus the $^{87}Sr/^{86}Sr$ isotope compositions. The plant data from this study are also plotted, along with published data from the primary glass production site of Tyre and the secondary glass production site of Banias, both in the Levant, together with the site of Ra's Al-Hadd in Oman (Leslie *et al.* 2006). The data for Tyre and Ra's Al-Hadd have been given arbitrary Sr concentrations of 30 ppm because these data are not available.

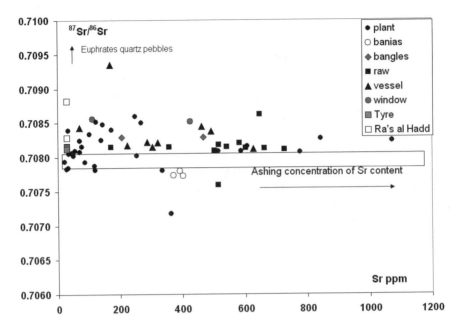

Fig. 3.4
Sr concentration versus $^{87}Sr/^{86}Sr$ isotope composition for 24 al-Raqqa glass samples and 31 plant ash samples compared with glass samples from other sites. The data for Tyre and Ra's Al-Hadd have been given arbitrary Sr concentrations of 30 ppm

There is a large spread of data with respect to Sr concentrations, and the majority of raw and vessel glass from this study fall on this trend. Banias data have a lower $^{87}Sr/^{86}Sr$ ratio consistent with production closer to Damascus or using Damascus-derived plants. The data from Tyre are tightly constrained isotopically and similar to the al-Raqqa data. One of the Ra's Al-Hadd samples displays a high $^{87}Sr/^{86}Sr$ isotope composition of 0.7139 and is beyond the range of any plant sampled in this study, a value that is explained (Leslie *et al*. 2006) by the possible contribution of a sand from an older continental source. Although this plot shows the effect of ashing on the Sr concentration and distinguishes Banias from the other sites, it is not a very good discriminant diagram for the different glass object types from al-Raqqa. It does however demonstrate that isotopic fractionation has *not* occurred during glass production.

NEODYMIUM

The silicate source composition of the glass has been classified using oxygen and neodymium isotopes. Two basic types of silica sources are possible: quartz pebbles, that are collected, crushed and melted, or a sand source. The latter can contain significant levels of other minerals, such as feldspar, that will affect its isotopic and chemical composition, whereas quartz pebbles provide a very clean source of silica.

Fig. 3.5a shows Nd isotope compositions of the glass artefact types plotted against Nd concentrations. The glass used to make different artefact types from al-Raqqa starts to separate out into groups. The raw glasses plot in an area with Nd levels of 4-7ppm and with $^{143}Nd/^{144}Nd$ between 0.51195 and 0.51210. One sample of raw glass (Raqqa 40) plots as a complete outlier, at 11 ppm and $^{143}Nd/^{144}Nd$ $c.$ 0.51237. Two window glasses and the two bangles from al-Raqqa have similar Nd concentrations to the raw glass, but are more radiogenic, with values of between 0.5122 and 0.51225. The window glass dates to the 9th century AD and the bangles to the early 11th century AD. The vessel glass defines an elongated field which is typical of a mixing curve formed by the samples being derived from differing proportions of two end members, one having a low Nd concentration and relatively high $^{143}Nd/^{144}Nd$ ratio and the other having higher Nd concentration and lower $^{143}Nd/^{144}Nd$. This suggests that a relatively pure quartz source was used from a relatively older geological source for one glass end member, whereas the other end member glass is formed from a less pure quartz source of a more radiogenic, and hence probably geologically younger, source.

Fig. 3.5b shows the Nd isotope signature of the samples according to their chemical composition plotted against Nd concentration. The types of chemical compositions are those defined in Henderson *et al.* (2004). Types 1, 2 and 4 are all plant ash glasses; type 3 is a natron glass so it has been excluded from consideration. Type 2 glasses are characterized by containing the highest concentrations of Nd, including the outlier glass Raqqa 40 (*ibid.* Table 1, analysis 22) which contains 4.88% MnO, contributing to its emerald green colour. Type 4 glasses can generally be distinguished by being more radiogenic that type 1, with two exceptions to this for both glass types. Moreover, three pairs of samples have almost identical Nd signatures.

Fig. 3.5a also shows that the twisted and plain bangle samples (Raqqa 268 and 269), three fragments of flasks (Raqqa 50, 58 and 81) and cobalt blue and emerald green window fragments form a group which is united by its Nd isotope ratios of between 0.512190 and 0512244; they were therefore made using a silica source differing from that used to make typical Raqqa glasses. Fig. 3.5b shows that the vessel samples are of compositional types 1, 2 and 4, whereas the bangle and window samples are all of type 4. This underlines the point that isotopic groups do not necessarily correlate with groups defined by chemical compositions.

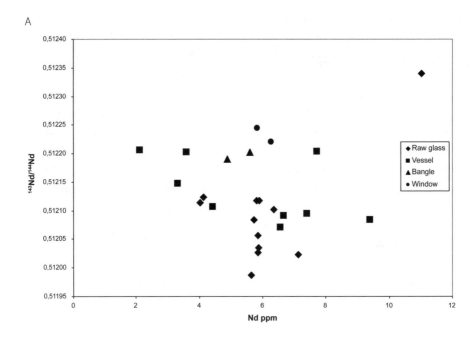

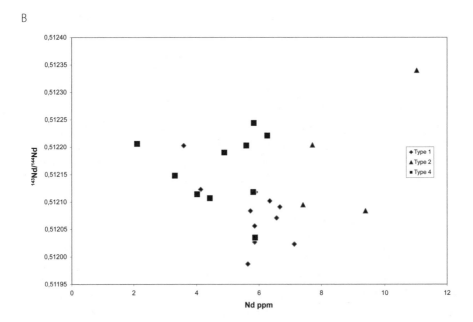

Fig. 3.5
Nd isotope composition of the glass artefacts from from al-Raqqa plotted against the Nd concentration for artefact types (a) and compositional types (b)

In Fig. 3.6 the Nd isotopes are plotted against the $\delta^{18}O_{vsmow}$ composition for each sample (Nd data for Banias, Ra's Al-Haad and Tyre are not available). This does not produce a particularly good discrimination between glass artefact types, and there appears to be little correlation between the two discriminants; for $\delta^{18}O_{vsmow}$ the majority of the al-Raqqa data fall between 13-14 per mil within the range of the Euphrates pebbles which gives 14.1 ± 3.3 (2σ, n = 13) an average value for all but two of the most extreme compositions (Henderson *et al.* 2005). The al-Raqqa data can however be distinguished from silica sources used to make Tyre glass together with two of the Banias samples and one sample from Ra's al-Hadd. This means that by using oxygen isotopes to characterize silica sources we are able to distinguish between al-Raqqa and Tyre, though there is some overlap with the Banias data.

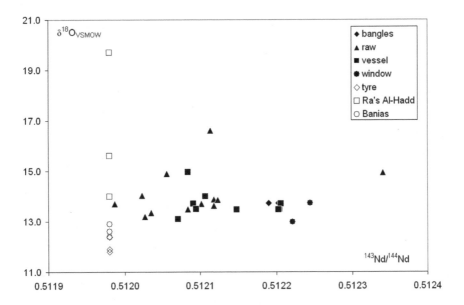

Fig. 3.6
Nd isotopic compositions plotted against $\delta^{18}O_{vsmow}$ compositions for al-Raqqa samples and other glass types (Nd data for Banias, Ra's Al-Haad and Tyre are unavailable)

Fig. 3.7a is the equivalent to a standard geological plot used to discriminate between geological sources of rocks and is usually plotted using an 'epsilon notation'. However, here measured ratios are used. In the geological system the behaviour of Sr and Nd isotopes is linked through magmatic and contamination processes resulting in trends and curves on such a diagram. Because of the de-coupled nature of the source of Sr and Nd in man-made glass, such trends are not expected or seen. However, sample Raqqa 40, a piece of raw emerald green glass, is interesting as it plots in an area of the diagram that is diagnostic of basaltic rocks and for both its Sr (in the Damascus area) and Nd composition. It is also distinguished by an elevated Nd concentration (11ppm) in comparison to that of the other glasses. This suggests that the glass was produced from a basic, geologically un-evolved (basalt) terrain where the sand was a direct derivative and the plants were grown on the same lithology, thereby transferring the correlated isotope signature of the rock directly into the glass. The plot also shows that two vessel glasses (Raqqa 47 and 54, both undecorated bowl fragments) group tightly with five raw glasses and can be considered to be al-Raqqa products; the samples are either of compositional type 1 or 4 (Fig. 3.7b). Six other raw glasses are distinct.

Discussion

A number of new isotopic results and archaeological interpretations are presented here. The chemical distinction between two principal types of plant ash glass found at al-Raqqa has been further largely confirmed by the isotopic work, with a small number of exceptions. Inland silica sources of different geological ages and purities were used to make the two contemporary glass types and they must have largely remained segregated on site. Perhaps their impurity levels gave the glass specific working properties, but whatever the explanation a decision had been made, as part of the *chaîne opératoire,* to keep them separate on site. There is, perhaps, a greater chance that silica, as opposed to plant ash, could be imported to production centres. However, if the production centre of plant ash glass is located close to the silica source, then it will provide a second means of provenancing the glass. If a distinctive silica source were used consistently, this would be a further means of characterising the glass. The Sr isotope results bolster our initial findings (Henderson *et al*. 2005) of a constrained field for plants growing in the vicinity of al-Raqqa. The ellipse of Sr results which characterizes al-Raqqa production in Fig. 3.4 encloses results for raw and vessel glass and for plants growing near al-Raqqa. This is the first time that this has been demonstrated for a plant ash glass production site. This proves that *local plants* were used to make the glass of *both* main compositional types of plant ash glass, and that no isotopic fractionation occurred. This forms the basis of provenance.

Mixing of glass during production may confuse provenance studies. On the basis of chemical analyses we suggested (Henderson *et al*. 2004) that mixing formed part of the experimental phase that occurred in 9th century AD al-Raqqa.

A

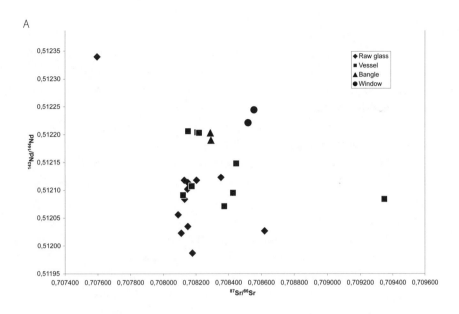

B

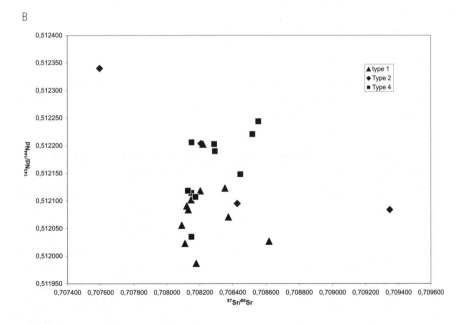

Fig. 3.7
Sr-Nd isotopic composition for al-Raqqa glass according to artefact types (A) and compositional types (B)

In archaeological terms this suggests that 'foreign' glass, that is glass that has been made using raw materials of different ages found at some distance from al-Raqqa, was mixed with raw al-Raqqa glass. If this was the case, the most likely interpretation of this would be that mixing of imported raw and vessel glass occurred in a crucible in beehive-shaped furnaces, prior to glassblowing. Interesting variations in Sr and Nd isotope signatures are seen in the 'experimental' (type 4) glass type (Henderson *et al.* 2004, type 4) found in al-Raqqa (Fig. 3.5b and 3.7b). Mixing lines for type 4 glass are visible in both Fig. 3.5b and 3.7b. On the left hand side of Fig. 3.5b there is a mixing line for five samples of type 4 glass (three vessels and two raw), the older samples with higher values being vessels. The evidence for mixing of type 4 glass is also substantiated by Fig. 3.7b: on both left and right hand sides three samples form mixing lines. It is notable that there is a constrained Nd/Sr isotope field for type 4 *raw* glass, with a wider variation in type 4 artefacts (Fig. 3.7a and 3.7b). A single type 4 vessel matches the raw glass. On the right hand side of Fig. 3.5b a mixing line is formed from two type 2 raw glasses at the extreme ends, with a single type 2 vessel sample in between. There is also a clear mixing line for the four type 2 glass samples in Fig. 3.7b stretching across the full range of Sr isotope ratio values. It was originally suggested on compositional grounds that type 2 glasses were a result of mixing (Henderson *et al.* 2004, 460).

The isotopic results have provided evidence of the import to al-Raqqa of vessels, armlets and window glass: a clear example of a vessel import is Raqqa 61 with a $^{87}Sr/^{86}Sr$ ratio value of 0.7093. This derived from the Princess's palace in al-Rafica and is of 12th century AD date, an indication that at least one other source of glass was used by this time. Three flasks from 9th century AD Tell Zujaj, two windows from the contemporary 9th century AD palace complexes in al-Raqqa and two bangles from early 11th century AD Tell Fukhkhar are united by their distinct Nd and Sr isotope ratios. This indicates firstly that they were imported and secondly that the same raw material source was used over some 200 years. A 9th century AD sample of *raw* glass that falls well outside the Sr field for al-Raqqa has a $^{87}Sr/^{86}Sr$ ratio value of 0.7087. The existence of this piece suggests that a small proportion of foreign raw glass found its way to the site. This can hardly be regarded as unexpected, given the political significance of al-Raqqa in the early Islamic period, but it was claerly unusual. As more isotopic results become available for ancient glasses it will become easier to interpret this result. The full archaeological potential of using isotopic analysis to provenance Islamic glass, including contributions to the evidence for local manufacture, the supply system and trade networks, is discussed elsewhere (Henderson *et al.* in press).

Conclusions

The use of Sr, Nd and O isotope analyses of plant ash glass provides a means of defining where in the Islamic world raw and vessel glass was made (provenance). There is some evidence from chemical analyses for the use of regionalized plant ash glass recipes (Henderson 2003). Nevertheless isotopic analysis has a clear means of cutting through some of the ambiguity provided by such evidence and even providing evidence for glass production where there is no direct archaeological evidence. By using the raw primary glass found at al-Raqqa as a basis for establishing both chemical and isotopic variations for a sufficiently large number of samples from a large Islamic production site for the first time, it has become possible to provide a yardstick against which to compare results from the rest of the Islamic world. The manufacture of raw glass ultimately controlled by caliphs and regional governors should be reflected in such isotopic signatures. We have contributed here to the 9th century AD evidence for a growing sophistication of what it meant to be Islamic, as reflected in the materials made and used. Before the 9th century AD Muslims were largely dependent on an essentially Byzantine glass technology.

The constrained field of Sr isotope ratio signatures for plants local to al-Raqqa, raw glass from al-Raqqa and some vessel fragments can be regarded as the 'al-Raqqa signature'. There is a clear link between the plants, the raw glass and the vessel fragments. It is shown here how important it is to carry out a statistically valid set of isotopic determinations, so that data for production sites can be defined, thereby making it easier to have confidence in the results for 'unknowns'.

The determination of Sr isotope variations across Syria and into the Lebanon has provided the first assessment of bioavailable Sr in the zone. There is evidence of a contrast between the quaternary limestone landscape of northern Syria and the younger geology of southern Syria dominated by basalt. This suggest that glass from Damascus, if made using plant ash derived from samples grown on basalt, will have a lower $^{87}Sr/^{86}Sr$ ratio than al-Raqqa glass and hence be distinguishable.

A limitation of using Sr isotope ratio determinations to provenance glass is however shown by the indistinguishable signatures for al-Raqqa, Syria, from the coastal sites of Tyre, Lebanon and Ugarit, Syria. It is therefore not necessary to interpret the discovery of such a similarity in Sr isotope signatures in glass from al-Raqqa and Tyre as being evidence for the export of plant ash from al-Raqqa to Tyre (Leslie *et al.* 2006, 262).

In general, the Nd isotope signatures in al-Raqqa glasses show a correlation with (chemical) compositional type, indicating that distinct silica sources were used and that two separate production procedures were followed in order to manufacture the two contemporary 9th century compositional types. There is no *a priori* reason why chemical and isotopic compositions should be correlated. Type 1 (plant ash) glass was made from a silica source with a less radiogenic (geologically older) Nd isotope signature (perhaps Euphrates sand) when compared with Type 4. Type 2 glass contains variable ratios and amongst the highest concentrations of Nd.

The isotope composition for bangles and window glass from al-Raqqa plot above the field of al-Raqqa raw glass, and hence neither is a primary product of the al-Raqqa site but were imported. The 11th century AD bangles and 9th century AD window glass are distinctive because they are of deep translucent colours. This may suggest, though on the basis of small data sets, that other production sites, where the bioavailable Sr was more radiogenic, specialized in the manufacture of highly coloured Islamic window glass and bangles.

There is clearly potential to use Sr isotopes as a means of provenancing Islamic plant ash glass, but over relatively broad areas and where there is sufficient contrast in the bedrock geology. It is possible that where Sr isotopes are unable to distinguish between production zones, the use of contrasting silica sources local to the production centres characterized by Nd and O isotopes may offer a way forward.

We have detected a sample of *raw* plant ash glass from al-Raqqa (Raqqa 40) with very distinct (younger) more radiogenic isotope signatures compared to the typical raw glasses found in al-Raqqa. Glass fused in Damascus to the south would produce such a signature. Therefore in spite of the apparent ease with which the glass could be melted locally, raw glass was occasionally imported to al-Raqqa from other production sites. The sample is of a deep translucent emerald green colour, so it is possible that it was imported from a site which specialized in making highly coloured glass. The isotopic signatures of highly coloured windows and bangles provide more evidence for this. A 12th century AD vessel glass fragment (Raqqa 61) with an anomalously high Sr/Sr ratio is an import: by this time a separate source of glass was being exploited.

We have detected some evidence for mixing/recycling of glass using isotopic data. The interpretation of chemical analyses suggested that mixing (experimentation) was involved in the 9th century AD manufacture of plant ash types 2 and 4. The mixing of both calcium and silica raw materials of different ages has clearly occurred. In the case of type 2 it is apparent that one end member raw glass sample (Raqqa 40) was fused from silica and plant ashes deriving from a geologically younger area.

Acknowledgements

We thank Hilary Sloane for providing the oxygen isotope data. JE publishes with permission of the Director of BGS. JH thanks the British Academy for a Research Readership and we thank Dr Lloyd Weeks for offering helpful comments on the paper. We thank the BGS for funding the isotopic determinations (IP/817/05-04).

References

F. Aldsworth, G. Haggarty, S. Jennings, D. Whitehouse, 2002, Medieval glassmaking at Tyre, Lebanon, Journal of Glass Studies, 44, 49-66.

A. Alexandre, I. Basile-Doelsch, C. Sonzogni, F. Sylvestre, C. Parron, J.D. Meunier, F. Colin, 2006, Oxygen isotope analyses of fine silica grains using laser-extraction technique: comparison with oxygen isotope data obtained from ion microprobe analyses and application to quartzite and silcrete cement investigation, Geochimica et Cosmochimica Acta, 70, 2827-2835.

I. Andreescu-Treadgold, J. Henderson, 2006, Glass from the Mosaics on the West wall of Torcello's Basilica, Arte Medievale, 2, 87-140.

K. Asch, 2005, IGME 5000 - 1:5 Million International Geological Map of Europe and Adjacent Areas - final version for the internet. BRG, Hannover.

E. Ashtor, 1992, Levantine alkali ashes and European industries, in: B.Z. Kedar (ed.) Technology, Industry and Trade: the Levant versus Europe, 1250-1500, Variorum, 475-522.

Y. Barkoudah, J. Henderson, 2006, Plant ashes from Syria and the manufacture of ancient glass: ethnographic and scientific aspects, Journal of Glass Studies, 48, 297-321.

G.F. Bass, 1984, The nature of the Serçe Limani glass, Journal of Glass Studies, 26, 64-69.

Z.R. Beydoun, 1977, The Levantine countries: the geology of Syria and Lebanon (maritime regions), in: A.E.M. Nairn, W.H. Kanes, F.G. Stehli (eds.) The ocean basins and margins, volume 4A, The eastern Mediterranean, Plenum Press, 319-353.

J.L. Birck, 1986, Precision K-Rb-Sr Isotopic Analysis - Application to Rb-Sr Chronology, Chemical Geology, 56, 73-83.

R.H. Brill, 1970, Lead and oxygen isotopes in ancient objects, Philosophical Transactions of the Royal Society of London, A.269, 143-164.

R.H. Brill, 1999, Chemical analyses of early glasses, Corning Museum of Glass.

R.H. Brill, T.K. Clayton, C.P. Stapleton, 1999, Oxygen isotope analysis of early glasses, in: R.H. Brill, Chemical analyses of early glasses, Corning Museum of Glass, 303-322.

S. Carboni, G. Lacerenza, D. Whitehouse, 2003, Glassmaking in Medieval Tyre: The Written Evidence, Journal of Glass Studies, 45, 139–149;

R.N. Clayton, T.K. Mayeda, 1963a, The use of bromine pentafluoride in the extraction of oxygen from the oxides and silicates for isotopic analysis, Geochemica et Cosmochemica Acta, 27, 43-52.

R.N. Clayton, T.K. Mayeda, 1963b, Oxygen isotope fractionation between diatomaceous silica and water, Geochemica et Cosmochimica Acta, 27, 1119-1125.

A.P. Dickin, 1995, Radiogenic Isotope Geology, Cambridge University Press.

P. Degryse, J. Schneider, U. Haack, V. Lauwers, J. Poblome, M. Waelkens, Ph. Muchez, 2006, Evidence for glass 'recycling' using Pb and Sr isotopic ratios and Sr-mixing lines: the case of early Byzantine Sagalassos, Journal of Archaeological Science, 33, 494-501.

J.A. Evans, S. Tatham, 2004, Defining 'local signature' in terms of Sr isotope composition using a tenth-twelfth century AD Anglo-Saxon population living on a Jurassic clay-carbonate terrain, Rutland, UK, in: K. Pye, D.J. Croft (eds.) Forensic Geoscience: principles, techniques and applications, Geological Society of London Special Publication, 232, 237-248.

D. Foy, 1996, BEY 002, Context 24: less verres, Bulletin d'Archéologie et d'Architecture Libanaises 1, 90-97.

D. Foy, 2000, Un atelier de verrier à Beyrouth au debit de la conquête Islamique, Syria, 77, 239-290.

I. C. Freestone, Y. Gorin-Rosen, M. J. Hughes, 2000, Primary glass from Israel and the production of glass in Late Antiquity and the Early Islamic period, in: M.D. Nenna (ed.) La route du verre. Ateliers primaires et secondaires du second millénaire avant J.C. au Moyen Age, Travaux de la Maison de l'Orient Méditerranéen, 33, TMO, 65-82.

I.C. Freestone, K. A. Leslie, M. Thirlwall, Y. Gorin-Rosen, 2003, Strontium isotopes in the investigation of early glass production: Byzantine and early Islamic glass from the Near East, Archaeometry, 45, 19-32.

Y. Gorin-Rosen, 2000, The ancient glass industry in Israel: summary of the finds and new discoveries, in: M.D. Nenna (ed.) La Route du Verre. Ateliers Primaires et Secondaires du Second Millénaire av. J.-C. au Moyen Âge, Travaux de le Maison de l'Orient Méditerranéen, 33, 49-64.

S. Heidemann, 2003, Die Geschichte von al-RaqqaAl-Raqqa/al-Rafiqa ein Uberblick, in: A. Becker, S. Heidemann (ed.) Al-RaqqaAl-Raqqa II. Die islamishe Stadt, Philipp von Zabern.

S. Heidemann, 2006, The history of the industrial and commercial area of 'Abbasid Al-RaqqaAl-Raqqa, called Al-RaqqaAl-Raqqa Al-Muhtariqa, Bulletin of SOAS, 69, 33-52.

J. Henderson, 1999, Archaeological and scientific evidence for the production of early Islamic glass in al-Raqqa, Syria, Levant 31, 225-240.

J. Henderson, 2000, The Science and Archaeology of Materials, Routledge.

J. Henderson, 2002, Tradition and experiment in 1st millennium AD glass production – the emergence of Early Islamic Glass technology in late antiquity, Accounts of Chemical Research, 35, 594-602.

J. Henderson, 2003, Glass trade and chemical analysis: a possible model for Islamic glass production, in: D Foy, M.D. Nenna (eds.) Échanges et commerce du verre dans le monde antique, Montagnac, 109-123.

J. Henderson, in press, The provenance of archaeological plant ash glasses, in: A. Shortland, I.C. Freestone, Th. Rehren (eds.) From mines to microscopes - Studies in honour of Mike Tite, University College London Press.

J. Henderson, J. McLoughlin, D. McPhail, 2004, Radical changes in Islamic glass technology: evidence for conservatism and experimentation with new glass recipes from early and middle Islamic Raqqa, Syria, Archaeometry, 46, 439-468.

J. Henderson, S. Challis, S. O'Hara, S. McLoughlin, A. Gardner, G. Priestnall, 2005a, Experiment and Innovation: Early Islamic Industry at al-RaqqaAl-Raqqa, Syria, Antiquity, 79, 130–145.

J. Henderson, J.A. Evans, H.J. Sloane, M.J. Leng, C. Doherty, 2005b, The use of oxygen, strontium and lead isotopes to provenance ancient glasses in the Middle East, Journal of Archaeological Science, 32, 665-673.

J. Henderson, J. Evans, Y. Barkoudah, in press, The roots of provenance: glass, plants and isotopes in the Islamic Middle East, Antiquity, 2009.

J.C. Lamm, 1929-30, Mittelalterliche Gläser und Steinschnittarbeiten aus dem Nahen Osten, Forschungen zur islamischen Kunst, 5, Berlin.

K.A. Leslie, I.C. Freestone, D. Lowry, M. Thirlwall, 2006, Provenance and technology of near Eastern glass: oxygen isotopes by laser fluorination as a compliment to Sr, Archaeometry, 48, 253-270.

J.M. McArthur, R.J. Howarth, T.R. Bailey, 2001, Strontium isotope stratigraphy: LOWESS version 3: Best fit to the marine Sr-isotope curve for 0-509 Ma and accompanying look- up table for deriving numerical age, Journal of Geology, 109, 155-170.

J. Montgomery, J.A. Evans, G. Wildman, 2006, [87]Sr/[86]Sr isotope composition of bottled British mineral waters for environmental and forensic purposes, Applied Geochemistry, 21, 1626-1634.

A. Sillen, G. Hall, R. Armstrong, 1998, [87]Sr/[86]Sr ratios in modern and fossil food webs of the Sterkfontein valley: implications for early hominid habitat preference, Geochimica et Cosmochimica Acta, 62, 2463-2478.

M.S. Tite, A. Shortland, Y. Maniatis, D. Kavoussanaki, S.A. Harris, 2006, The composition of the soda-rich and mixed alkali plant ashes used in the production of glass, Journal of Archaeological Science, 33, 1284-1292.

K.H. Wedepohl, A. Baumann, 2000, The use of marine molluscan shells in the Roman glass and local raw glass production in the Eifel area (Western Germany), Naturwissenschaften, 87, 129-132.

The implications of lead isotope analysis for the source of pigments in Late Bronze Age Egyptian vitreous materials

A.J. Shortland

Introduction

Lead isotope analysis (LIA) has been one of the most widely used of the radiogenic isotope provenancing techniques in archaeology. It has achieved notable successes, particularly in the Mediterranean where it has been extensively used to provenance copper tools and ingots (see review, Stos in press). It has proved a controversial technique (for a review see Pollard in press), with questions raised about the grouping of ore bodies into fields and questions of mixing, especially in the reuse of copper waste. It is interesting to note that the rise of other isotopic provenancing techniques, for example Sr and Nd, has yet to have the level of scrutiny that has been applied to the Pb isotope system, although such techniques have many of the same strengths and suffer from the same weaknesses. Pb isotopes seem to have recently undergone a rebirth alongside these more novel isotopes and are now being applied once again. Certainly, given the right archaeological question and geological circumstances, the technique has great value.

Against the amount of effort and therefore analyses applied to metal objects, glaze and glass have received comparatively little attention, although lead ores and lead-based products from Egypt have been worked on by numerous groups (Brill *et al.* 1974, Stos-Gale and Gale 1981, Lilyquist and Brill 1993). Thermal Ionisation Mass Spectrometry (TIMS) has been the principal technique used to identify the ore bodies from which the lead has been obtained. The Isotrace Unit at the University of Oxford carried out the analyses published in Shortland (2006) and in Fig. 4.1 here, following the methodology laid out elsewhere (Gale and Stos Gale 1991). This paper takes the data already published previously (Shortland 2006) and discusses in more detail the implications of those results and the possible interpretations.

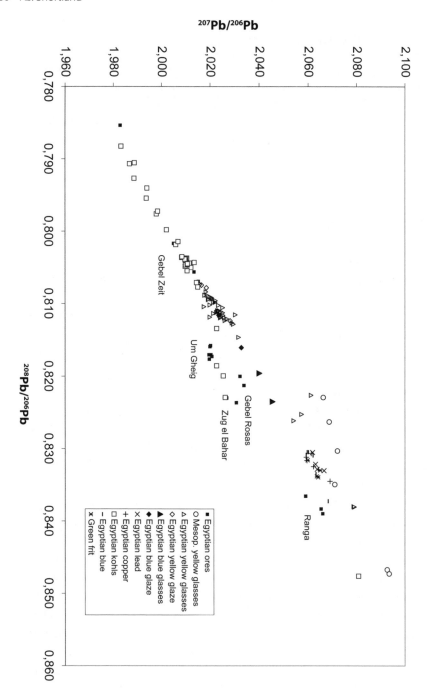

Fig. 4.1
Plot of ^{208}Pb/^{206}Pb against ^{207}Pb/^{206}Pb for the Egyptian and Mesopotamian materials comparing them to published results for lead ores (Stos-Gale and Gale 1981, Brill *et al.* 1974, Pouit and Marcoux 1990, Lilyquist and Brill 1993)

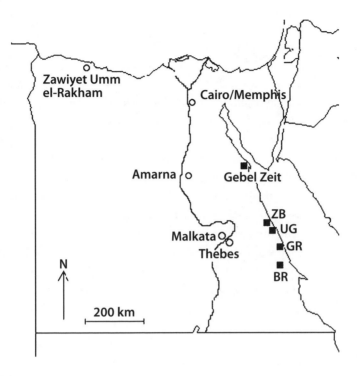

Fig. 4.2
Map of Egypt showing localities mentioned in the text. BR = Bir Ranga; GR = Gebel el Rosas;
UG = Umm Gheig; ZB = Zug el Bahar

Results

Shortland (2006) has analyses of galena kohls, lead weights, copper objects and
vitreous materials in the form of Egyptian blue pigment, faience glazes and glass.
Most of these date to a relatively tight time period during the 14th century BC and
came from both Mesopotamian and Egyptian sites. Comparative Egyptian lead
ore data were difficult to acquire, since the ores occur in significant quantities
only along the Red Sea coast of Egypt (Fig. 4.2). Although they are still mined
commercially at several large mining facilities (Said 1990, 543–550), they lie
within a zone of security sensitivity for the Egyptian government and so it is
difficult for even Egyptian geologists, let alone westerners, to visit the area. Despite
several petitions from the author to visit key mines along the coast this has proved
impossible. The ore bodies are (from south to north, Fig. 4.2): Bir Ranga (24.4°N),
Gebel el Rosas (25.2°N), Umm Gheig (25.7°N) Zug el Bahar (25.9°N) and Gebel
Zeit (27.9°N). The small numbers of lead isotope analyses already published for
these mines (El Goresy *et al*. 1998, Stos-Gale and Gale 1981, Brill *et al*. 1974,

Pouit and Marcoux 1990, Lilyquist and Brill 1993, 59–65, Castel *et al.* 1988, Castel and Soukiassian 1989) are therefore all that is available, and show only the general isotopic area in which the mines might plot. This represents nowhere near enough analyses to determine the size or shape of the lead isotope fields for each of the mines. In order to supplement this small number of LIAs for Egyptian ore sources, a second body of material was analysed. Egyptians made wide use of kohl, or black eye paint. Analyses of kohls contemporary with the glass and others have shown that the great majority of them were galena, the principal lead ore (Hassan and Hassan 1981; Lucas and Harris 1962, 80–83). Shortland (2006) subjected thirty-two of these kohls to LIA. The range of their values can be taken as representing a lead ore body to which the Egyptians had access. In fact some evidence for ancient mines for the exploitation of the lead ores for kohl is found at Gebel Zeit, dating from the 12th Dynasty (1985–1795 BC) to the New Kingdom (1550–1069 BC) and potentially beyond (Castel and Soukiassian 1989). Gebel Zeit is by far the best studied of all the ancient lead mining sites in Egypt, and the conditions and methods used in the extraction of lead ores in antiquity are amply demonstrated at the site (Castel and Soukiassian 1989). Two major sites have been recognized at Gebel Zeit, Site 1 and Site 2, separated by some 4 km and each with apparently different lead isotope characteristics, although only two values are available for each. Two of the kohls have a $^{207}Pb/^{206}Pb$ of around 0.807, very close to the Gebel Zeit Site 2 ores, whereas fifteen are between 0.800 and 0.808 and plot between the two Gebel Zeit sites. A further eleven kohls plot near or between the two Site 1 values. The four LIAs for these two sites match well with most of the kohls, suggesting that this may have been the source for the kohls examined (see Fig. 4.1).

EGYPTIAN LEAD METAL, COPPER METAL AND COPPER ALLOYS

Four samples of lead from fishing weights from the Amarna site were analysed in Shortland (2006) and found to have a $^{207}Pb/^{206}Pb$ of 0.830-0.835. They do not plot close to the fields for any of the Egyptian Red Sea ores, suggesting that these ores were not used to make the lead in the weights. This supports earlier studies (Stos-Gale and Gale 1981) which have shown that no lead object analysed so far appears to have been made from any of the characterized Red Sea ore sources. The source of lead metal for Egypt therefore seems to lie outside the country. The copper metal and copper alloys have very similar lead isotopic characteristics to the lead metals and are therefore also interpreted as being from a source outside Egypt.

PIGMENTS: EGYPTIAN BLUE AND GREEN FRIT

Egyptian blue and green frits are high copper vitreous materials, typically around 15–20% CuO, an order of magnitude higher than the contents of the glasses. During the New Kingdom, they had no lead deliberately added (Hatton 2003), so the trace levels of lead in the pigment therefore derive almost exclusively from the copper, where it is present as a natural part of the copper ore and therefore metal. If the same copper being used locally in metalworking is used in the frits, then the lead isotopic characteristics should be identical. Fig. 4.1 shows that for Egyptian blue and green frits this is indeed the case, showing that the copper used in their manufacture and the copper used for tools and weapons in Egypt have the same source, outside Egypt.

GLASSES

Twenty-nine LIA results for glasses are shown in Fig. 4.1 alongside seventeen LIAs of glass published by Lilyquist and Brill (1993, Table 2). The yellow glasses from Egypt can be divided into two groups. The main group, Group A, has a $^{207}Pb/^{206}Pb$ in a fairly tight cluster of between 0.808 and 0.815. They are close to the Gebel Zeit Site 2 ores and lie along their trendline, but with slightly higher ratios. The second 'group', Group B, is a much looser scatter, with a higher $^{207}Pb/^{206}Pb$. In this are all the early glasses from the reign of Tuthmosis III (1479–1425 BC), a single glass rod from Amarna and the four yellow glasses from Mesopotamia.

The two blue, copper-coloured glasses are distinct from the yellow glasses as one might expect, since their trace levels of lead are theoretically derived from low levels of lead in the copper colorant, exactly as with the frits above. The values for the blue glasses fall between the two yellow glass Groups A and B, well away from any other material, with a $^{207}Pb/^{206}Pb$ of around 0.820-0.825.

FAIENCE

Four samples of yellow faience, have identical values to some of the kohls and one of the ores from Gebel Zeit Site 1 and are close to the Egyptian yellow glass Group A. The single piece of green faience plots within the Group A glasses, but at the high $^{207}Pb/^{206}Pb$ end of the Group. The two blue faience pieces fall either within or close to Group A, with lower $^{207}Pb/^{206}Pb$ and $^{208}Pb/^{206}Pb$ than the blue glasses, echoing the pattern seen in their yellow equivalents.

Discussion

Insufficient ore body LIAs exist for Egyptian ores to make statements of certainty as to the sources of Egyptian vitreous materials. However, some indications can be given. Firstly, the Egyptian ores have very low $^{207}Pb/^{206}Pb$ and $^{208}Pb/^{206}Pb$ values, lower than almost all other ore bodies known or suspected to have been used in the ancient world. This means that they are clearly distinct from other major ore sources and therefore free of the complexities of interpretation that the more crowded areas with higher $^{207}Pb/^{206}Pb$ and $^{208}Pb/^{206}Pb$ present. In addition, the kohls seem to be a good indirect proxy for LIAs on galena direct from the ores and fall within the isotopic range of the Gebel Zeit ore body, known to have been exploited during the time at which the vitreous materials in the study were being produced. There is therefore good reason to work from the assumption that the kohls came from an Egyptian ore body, probably Gebel Zeit, and can be thought of as representative of that body. It is possible that the kohls represent multiple ore bodies mixed together, but since the majority of the kohls fall within the known range for Gebel Zeit this is perhaps less likely. Thus it is possible to speculate reasonably what the lead isotope characteristics of the Gebel Zeit ore body may be, and thus work from this premise.

It is clear that the lead isotope characteristics of the ores/kohls are very different from those of the lead and copper/copper alloy metals. The metals have similar lead isotope characteristics, with a $^{207}Pb/^{206}Pb$ of 0.830-0.840. Neither matches or is close to any known Egyptian source. Thus, although the lead ores of the Red Sea Coast are being exploited for the production of kohl, these same ores are not being smelted into metal. Why this may be is a very difficult question. It is possible that Egyptian ores are being made into metal and that metal is yet to be analysed, since relatively few lead metal objects are known and very few of these have been subjected to LIA. However, it is perhaps more likely that we see a division of production between metals and cosmetics, with the mines being specifically used for kohl and perhaps tasked to produce such by the major temples such as Karnak with which the mines have been linked. Thus lead appears to be imported from outside Egypt, although where this site may be is uncertain. It has been widely discussed elsewhere (Gale 1991; Stos-Gale *et al*. 1995), with the consensus of opinion being that a number of sources are possible, with Lavrion, Cyprus and the Taurus Mountains all supplying some metal.

A similar pattern can be described for copper, where, once again a non-Egyptian but otherwise uncertain source must be proposed (Stos-Gale *et al*. 1995). This is perhaps less surprising than with the lead, for although Egypt is rich in major lead sources, some of which are known to have been exploited, it is comparatively poor in copper mines. Some do exist, for example at Serabit el Khadim in the Sinai, where turquoise and other copper ores were mined, but these were not sufficient to supply Egypt's very large demand for the metal. Importing from major suppliers abroad therefore fits the expected model.

It is on close examination of the LIAs for the vitreous materials that the situation becomes more complex. Of them all, yellow faience seems the easiest to interpret since its analyses fall together within the area of the kohls and the Gebel Zeit Site 1 ores. It is likely, therefore that the lead in the lead antimonate which colours the yellow faience glaze was made from galena that came from Gebel Zeit. Thus the galena for the faience was being taken from the production of kohl, certainly suggesting a link between the two manufacturing communities. The early yellow glasses of Group B from the reign of Tuthmosis III have similar lead isotope characteristics to Mesopotamian yellow glasses, as shown by Lilyquist and Brill (1993, 61–62), but they are very variable, much more so than in the later Egyptian glasses. The conclusion has therefore been drawn that the early Egyptian yellow glasses were imported to Egypt from Mesopotamia (Lilyquist and Brill 1993, 62), which fits textual and archaeological evidence that this was the case (Shortland 2006). The later glasses with one exception all have a $^{207}Pb/^{206}Pb$ of 0.808-0.815 (Group A). However, they are not identical since they have higher $^{207}Pb/^{206}Pb$ values and they are also higher than the Gebel Zeit ores/kohls, although on their trendline. The answer to this problem may be indicated by the single later glass which does not fit within the low, Group A isotopic values. This is a yellow glass rod from the glassworking/making site of Amarna. This rod has a value closer to the early glasses, but is securely dated much later. As such it may represent 'old stock', preserved accidentally within the glassworking kit of the artisan, perhaps passed down from father to son over the one or two generations involved. Alternatively, it may indicate that, perhaps rarely, some glass with the higher Group B isotopic values continues to be imported into Egypt during the Amarna and perhaps into later periods. If this is the case, then the glass batch would be slightly contaminated by these relatively rare foreign pieces of glass; thus their lead isotope values would be higher, closer to those of the Mesopotamian glasses.

The yellow faience shows values close to the ores, suggesting no contamination. This may be because a faience glaze is produced from the raw materials on the faience body, so cannot be mixed with other faience glazes as glasses can be mixed. Unless foreign lead antimonate colorant is being imported as a raw material, yellow faience will retain the Egyptian lead isotopic signature. Thus several conclusions can be drawn: firstly, that the yellow lead antimonate colorant is produced in Egypt, and none appears to be imported (or the faience would show contamination too); secondly, that the glasses are being mixed from different sources giving a low level of contamination in them. It is possible to make a very rough prediction of what level of contamination might be seen by taking the average of the faience values (assumed uncontaminated pure Egyptian lead) and the average of the Mesopotamian values (assumed purely Mesopotamian lead) and calculating where the Egyptian yellow glasses lie on the mixing line between them. This gives a mixture of very roughly ten parts Egyptian to one part Mesopotamian. Thus one foreign yellow glass rod in the twelve mostly Egyptian-sourced yellow glasses is very approximately what might be expected.

The copper-blue vitreous materials are more complex still. The lead in them should be derived from the copper, and thus they should have similar lead isotope values to the copper objects analysed. This is the case with the pigments, which elegantly illustrate that this works and there is no reason why other copper-based vitreous materials should not do likewise. However, this is simply not the case. Neither the copper blue glasses nor the glazes have values similar to those of the pigments and metals – they both have significantly lower $^{207}Pb/^{206}Pb$. Why blue vitreous materials do not match the copper metals is not clear. Shortland (2006) speculated that the reason for the differences in lead isotope characteristics between the copper objects and copper blue glazes and glasses may be down to contamination in the kiln during melting and working. If the glass and glaze

Sample	UPP4	UPP5	UPP6	UPP22	UPP23	UPP24	UPP25	Average	UPP26	UPP27	UPP28	UPP29	UPP38	UPP39	Average
Colour	clls	clls	clls	clls	clls	clls	clls	clls	blue	blue	blue	blue	blue	blue	blue
Mn	121	153	108	302	4146	114	117	723	116	289	130	152	7575	532	2915
Fe	1596	4403	3926	2127	6121	1398	1417	2998	2049	6281	1986	2112	2415	3823	8530
Co	2,5	3,8	1,7	1,3	8,9	2,8	2,9	3,0	1,6	65,1	1,4	2,7	22,4	633	151
Ni	3,1	7,2	4,1	8,9	12,9	3,2	3,4	6,0	4,5	38,5	3,6	6,2	18,7	414	107
Cu	26	31	43	13	62	33	35	35	4212	15151	3967	4846	11564	1096	8222
Zn	11,2	19,2	30,9	15,1	147	22,9	23,1	39	122	81,6	12,4	20,5	178	681	281
As	0,4	0,7	1,3	1,8	3,0	0,5	0,4	1,2	17,9	55,4	6,6	33,4	28,4	12,9	33
Ag	0,44	0,08	0,14	0,04	0,42	0,10	0,11	0,19	0,56	2,46	0,18	0,40	2,27	0,84	1,64
Sn	11,1	10,5	12,9	7,9	7,4	11,8	12,0	10	380	1026	168	1712	794	102	853
Sb	7,3	8,1	1,7	<0.01	79,3	0,5	0,5	14,0	1133	477	15,5	6336	589	2533	2239
Au	<0.02	0,03	0,02	0,01	0,02	<0,02	<0,02	0,02	0,18	7,27	0,10	0,10	9,7	1,5	3,8
Pb	4,5	6,7	4,8	4,8	22,9	173	179	56	2903	39,4	7,0	11,6	114	50	715
Bi	<0.01	<0.01	0,06	0,06	0,02	0,04	0,31	0,07	0,18	0,12	0,13	0,31	0,09	0,05	0,29
Th	0,47	1,03	0,99	1,36	0,57	0,36	0,40	0,74	0,50	1,10	0,64	0,49	0,65	1,43	2,14
U	0,06	0,32	0,31	<0.04	0,49	0,14	0,17	0,21	0,20	0,58	0,22	0,18	0,40	0,79	0,82

Table 4.1
Results of LA-ICP-MS analyses for Egyptian colourless and blue glasses in this study

were in a kiln next to lead antimonite yellow glass or glazes, then it is possible that some of the lead could volatilise and become incorporated into the copper blue glaze, thus altering the lead isotope values of the glass/glaze. It is certainly true that the copper glasses have low lead contents, and in Shortland (2006) an average value of 35 ppm Pb was quoted. This was from a limited number of LA-ICP-MS analyses which formed part of an ongoing study when this paper was written in 2005. It was compared to colourless glass, which averaged 7 ppm Pb. The proposal in Shortland (2006) was that the 7 ppm Pb in the colourless glasses possibly showed that this volatilisation was happening, since this is orders of magnitude above the level of lead that would be expected in normal sand and plant ash. However, a major LA-ICP-MS study of several hundred LBA glasses has now been completed (Shortland, in preparation), and this has shown that these simple averages do not represent the whole story. A selection of elements from some of the colourless and blue glasses from this major study (including UPP27 and UPP39, the two blue glasses shown in Fig. 4.1) is shown in Table 4.1. The results in the table do not represent typical values, but rather show some of the range of values that exist in both copper blue and colourless glass. Examining the analyses of the colourless glasses, it is clear that some elements, for example Co, Ni and Zn, are reasonably consistent throughout, and the average represents a fair value for the composition of the glass. However, other elements, most significantly Pb, are much more variable, the Pb content ranging from 4.8 ppm to 179 ppm, the average in this case being very much less useful. The situation is even more variable in the blue glasses, where almost all the elements either used as colorants or associated with those elements used as colorants show variation much greater than that in the colourless glasses. This is especially true of Pb, which varies from 7 ppm to 2903 ppm. There could be several explanations for this which might also shed light on the curious LIA results for blue glasses and glazes. Firstly, as proposed by Shortland (2006), it could be that this is contamination in the kiln environment, with some glasses ending up more contaminated than others. Secondly, it could be that not only is the copper colorant bringing in lead to the system but perhaps there is another source. There is a strong possibility that the Sb, used in copper blue glasses as an opacifier, adds some trace (but significant) amounts of Pb. This fits in with studies of Sb opacifiers that have been carried out (see Shortland 2002). Thirdly, recent changes in ideas about how the glasses were made may be the answer. Most classical interpretations of the production of glass in this period involve just one stage of glassmaking, with or without a fritting stage. In other words, coloured glass is produced effectively in one firing, straight from the raw materials. However, recent work in several areas, but particularly at the glassmaking workshop at Qantir (Rehren and Pusch 2005), and some new work on glassmaking texts (Shortland 2007 and in press) has shown that it is more likely that glass was made in a two-stage process, first the manufacture of colourless glass, and then the colouring of that glass. Thus the colorant has to be stored and then added in some form. It is possible that it is in the storage and use of these colorants that cross-contamination occurs, rather than in the kiln itself. Finally, there is one more complicating factor, this time involving the Au content of the glasses. The maximum Au content of the colourless glasses shown is 0.02 ppm, which is roughly what might be expected. However, the minimum Au content of the blue glasses is 0.10 ppm and they mostly contain much more than this, averaging 3.8 ppm. Perhaps this once again represents contamination in the workshop, where gold may have been used for mounting the glass and as gold leaf over and around it. However, all these glass fragments are glass rods, so they have yet to be worked as finished objects and have probably had very little contact with gold. Perhaps the best explanation for the Au content involves the LBA perception of the value of glass. Glass was a prized commodity, valued as a semi-precious stone (Nicholson 1993), and blue glass especially, in that it was believed to have life- and health-giving properties (Robson 2001). Much ritual was attached to its manufacture (Shortland 2007) and presumably its use (Robson 2001). It is therefore possible, indeed even probable, that certain aspects of the glass manufacturing process, and even some raw materials, were added for ritual rather than any strictly functional reasons. The gold may have been one of these, added in very small quantities as perhaps an offering during the alchemical procedure of glassmaking. Similarly, if gold was added, it is possible that other things might have been too, even some containing lead traces. This would again complicate the pattern seen in the LIAs.

Thus there are at least four explanations for the curious LIA results with the copper coloured glasses and it is difficult to determine which may be the most important. It is certainly the case that several could be involved in the glasses. What is certain is that the trace element analyses provided by LA-ICP-MS give a valuable insight into the interpretation of the LIA results, which would be even more difficult to interpret without them.

Conclusions

Lead isotopic analysis has produced some very interesting results for Egyptian vitreous materials. Although ore body data are very limited, the use of kohls as a proxy for further data has given as good as possible an idea of the range of values present in one Egyptian ore body, probably that of Gebel Zeit on the Red

Sea coast (Castel and Soukiassian 1989). Extraction of galena for kohl seems to have been the main function of the many ancient mines exploiting this ore body. However, this galena was also used in the manufacture of the pigment lead antimonate yellow. This lead antimonate was used directly in yellow faience, which has lead isotope ratios practically identical to those of the Gebel Zeit ores. It was also used in yellow glasses of the Amarna period. The earlier glasses seem to have Mesopotamian lead isotope characteristics and were therefore probably imports from this area. These imports may rarely have continued into the later period where, in mixing with the Egyptian lead isotope characterised yellow glass, they produce a contaminated glass with LIA results slightly higher than those of the faience. One example of such an imported glass was found among the yellow glass rods from Amarna.

The copper metal in Egypt also seems to be imported, as has been stated many times in the past (Stos Gale *et al*. 1995). The copper rich Egyptian blue and green frits have identical values to the copper metal, showing that they were made from copper from the same source. However, the copper coloured blue glazes and glasses have curious values which are hard to interpret. Using LA-ICP-MS analyses to give trace element data on the glasses shows how complicated they can be. It seems that the Pb content of the glasses is very variable, and where this lead is coming from is far from clear. It may be contamination in the kiln, additional lead brought in with the antimonite opacifier, contamination of a colorant phase added to the glass, or deliberate ritual addition of one or more components containing lead – indeed several of these could be happening at once. This creates a very complex pattern for LIA interpretation in blue glasses and glazes, where, unlike with yellow glasses, their low levels of lead make contamination effects very significant. It is clear that the use of LA-ICP-MS with LIA gives a valuable insight into the complexity of the systems involved and is a useful aid in interpretation.

Acknowledgements

The author would like to thank the many museums and individuals who have given samples for this project. It would not have been possible without Helen Whitehouse and the Ashmolean Museum, Oxford, who contributed the bulk of the kohl samples and a great deal of time to allow their analysis by XRF and LIA. Museum collections in the Victoria Museum of Egyptian Antiquities (Uppsala), Nationalmuseet (Copenhagen), National Museum of Scotland, Petrie Museum of Egyptian Archaeology (UCL) and the Liverpool University Museum also contributed. Liverpool University also lent samples of Egyptian blue from their excavation at Zawiyet Umm el-Rakham through Steven Snape, the excavation director. The analysis of these Egyptian blue samples was coordinated by Gareth Hatton. Joan Oates of the McDonald Institute, Cambridge, contributed samples from the Tell Brak excavation collections which provided a valuable comparison for the Egyptian glasses. The author would like to thank Sophie Stos and the Staff of the Isotrace Laboratory in Oxford for their analytical expertise.

References

R.H. Brill, I.L. Barnes, B. Adams, 1974, Lead isotopes in some Egyptian objects, in: Bishay, A. (ed.) Recent Advances in the science and technology of materials, Plenum Press.

G. Castel, G. Pouit, G. Soukiassian, 1988, Les mines de galene pharaonique du Gebel Zeit (Egypte), dans le Miocène du rift de la Mer Rouge, Chronique de la Recherche Minière, 492, 19-32.

G. Castel, G. Soukiassian, 1989, Gebel el-Zeit - Les mines de galene (Egypte, IIe millenaire av. J.-C.), volume 1, Fouilles d'Institut Francais de Archeologie Orientale.

A. El Goresy, F. Tera, B. Schlick-Nolte, E. Pernicka, 1998, Chemistry and Lead Isotopic Compositions of Glass from a Ramesside Workshop at Lisht and Egyptian Lead Ores: A Test for a Genetic Link and for the Source of Glass, in: C.J. Eyre (ed.) Proceedings of the Seventh International Congress of Egyptologists, Orientalia Lovaniensia Analecta 82, 471-481.

N.H. Gale, 1991, Copper oxhide ingots: their origin and their place in the Bronze Age metals trade in the Mediterranean, in: N.H.Gale (ed.) Bronze Age Trade in the Mediterranean, SIMA, 90, Paul Astroms Forlag, 197-239.

N.H. Gale, Z. Stos Gale, 1991, Lead isotope studies in the Aegean, in: A.M. Pollard (ed.) Advances in science based archaeology, Royal Society/British Academy Special Publication, 63-108.

A.A. Hassan, F.A. Hassan, 1981, Source of galena in Predynastic Egypt at Nagada, Archaeometry, 23, 77-82.

G.D. Hatton, A.J. Shortland, M. Tite, 2003, Egyptian Blue: Where, When, How?, in: R. Ives, D. Lines, C. Naunton, N. Wahlberg (eds.) Current Research in Egyptology III, BAR International Series, 1192, 35-44.

C. Lilyquist, R.H. Brill, 1993, Studies in ancient Egyptian glass, Metropolitan Museum of Art.

A. Lucas, J.R. Harris, 1962, Ancient Egyptian materials and industries, 4th Edition.

A.M. Pollard, in press, What a long strange trip it's been: lead isotopes and archaeology, in: A.J. Shortland, Th. Rehren, I.C. Freestone (eds.) From mine to microscope; Studies in honour of Mike Tite, University College London Press.

G. Pouit, E. Marcoux, 1990, Les mineralisations Pb-Zn de la couverture miocene de la bordure du rift de la mer Rouge, rapports de la geochimie isotopique du plomb.

T. Rehren, E.B. Pusch, 2005, Late Bronze Age Glass Production at Qantir-Piramesses, Egypt, Science, 308, 1756-1758.

E. Robson, 2001, Society and Technology in the Late Bronze Age, a tour of cuneiform sources, in: A.J. Shortland (ed.) Social context of technological change, 39-58.

R. Said, 1990, The Geology of Egypt, Brookfield VT.

A.J. Shortland, 2002, The use of antimonate colorants in early Egyptian glass, Archaeometry, 44.4, 517-531.

A.J. Shortland, 2006, The application of lead isotopes to a wide range of Late Bronze Age Egyptian materials, Archaeometry, 48, 657-671.

A.J. Shortland, 2007, Who were the Glassmakers? Status, Theory and Method in Mid-Second Millennium Glass Production, Oxford Journal of Archaeology, 26, 261-274.

A.J. Shortland, in preparation, The analysis of second millennium BC glass from Egypt and Mesopotamia, part 2: new LA-ICPMS analyses, Archaeometry.

Z. Stos, in press, Across the wine dark seas... sailor tinkers and royal cargoes in the late Bronze Age Eastern Mediterranean, in: A.J. Shortland, Th. Rehren, I.C. Freestone (eds.) From mine to microscope, Studies in honour of Mike Tite, University College London Press.

Z. Stos-Gale, N.H. Gale, 1981, Sources of galena, lead and silver in Predynastic Egypt, Actes du XXème Symposium International d'Archéometrie, Revue d'Archéometrie, 5, 285-295.

Z. Stos-Gale, N.H. Gale, J. Houghton, 1995, The origin of copper metal excavated in El Amarna, in: W.V. Davies, L. Schofield (eds.) Egypt, the Aegean and the Levant: Interconnections in the 2nd millenium BC, British Museum Press, 127-135.

Kelp in historic glass: the application of strontium isotope analysis

David Dungworth, Patrick Degryse, Jens Schneider

Introduction

This paper explores the nature of strontium isotopic variation in general and argues that some provenance studies based on strontium isotopes are not as secure as they may appear. It is further argued that strontium content and isotopic ratios can make a significant contribution to our understanding of one particular type of raw material used in glass manufacture: kelp (seaweed).

Strontium isotopic ratios in nature and their use in geology and related disciplines

There are four naturally occurring isotopes of strontium (Table 5.1), of which only ^{87}Sr is radiogenic, being produced by the decay of ^{87}Rb. The proportion of ^{87}Sr in a rock (measured as the ratio of ^{87}Sr to ^{86}Sr) will increase over time depending on the age of the rock and its rubidium concentration. This phenomenon was used in the 1960s to develop a method of dating igneous rocks (e.g. Moorbath and Bell 1965; see Dickin 2005; Faure and Powell 1972; Faure and Mensing 2005 for further details). The use of strontium isotopes in this way led to the accumulation of fairly large data sets, but these tended to concentrate on geological deposits where dating was a major research theme. Thus, there are hundreds of analyses of igneous rock in Scotland but only a handful of results for England.

Isotope	Occurrence
^{84}Sr	0.56%
^{86}Sr	9.86%
^{87}Sr	6.99%
^{88}Sr	82.59%

Table 5.1
The four naturally-occurring isotopes of strontium

The use of strontium isotopes as a method of determining the age of a rock has been criticized as it relies on a number of assumptions, the most crucial being that rocks have been closed systems since their formation. Unfortunately few rocks satisfy this condition; they can receive and lose material over geological time, and this can have a significant effect on the apparent age as determined by strontium isotope analysis. Nevertheless, different geological regions will often have different strontium isotope ratios, and this has been used to trace catchment hydrology or to track the sources of water (Kendall *et al.* 1995).

Although the water in different rivers often has different strontium isotope ratios (reflecting the catchment geologies), the ratios in modern seawater are extremely uniform (cf. Burke *et al.* 1982). This uniformity is due to the large volume of water in the world's oceans and their rapid mixing. The negligible Rb content of most marine carbonates means that they have not undergone ^{87}Sr enrichment, and their strontium isotope ratios reflect the conditions under which they were formed. This has provided a means of determining the changes in the strontium isotope ratio over geological time (Fig. 5.1), which has provided information about global tectonic processes. In particular, the rapid rise in strontium isotope ratios over the last 40 million years has been related to dramatic mountain building and erosion activity in areas such as the Himalayas (Dickin 2005, 63–65). The latest strontium isotope curve (Fig. 5.1) is based on the careful selection of samples (often discrete fossils within marine carbonate rocks) and privileges 'less scattered data' (McArthur *et al.* 2001, 158). The data which have been excluded (cf. Burke *et al.* 1982) diverged from the data in Fig. 5.1 due to analytical/calibration problems with early data and the nature of the samples collected. Some samples are likely to have come from deposits which contained traces of rubidium, which will have elevated the $^{87}Sr/^{86}Sr$ ratio.

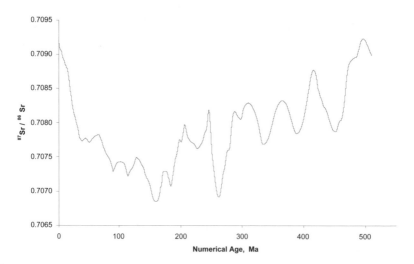

Fig. 5.1
Strontium isotope ratios of marine carbonate rocks (after McArthur *et al.* 2001)

Strontium isotope analysis of skeletal material

The fact that strontium isotopes are usually a reflection of underlying geology has long been recognized as a potential tool in archaeological studies (e.g. Stos-Gale 1989, Table 12.1). There are two main fields where the use of strontium isotopes is being applied to archaeological problems: skeletal material (Bentley and Knipper 2005, Bentley *et al.* 2003, Bentley *et al.* 2004, Budd *et al.* 2004, Ericson 1985, Chenery *et al.* 2006, Stoodley *et al.* 2006, Grupe *et al.* 1997, Hodell *et al.* 2004, Montgomery *et al.* 2003, Price *et al.* 2002, Price *et al.* 1994, Price *et al.* 2004, Schweissing and Grupe 2003, Sealy *et al.* 1991) and vitreous materials (Degryse *et al.* 2006, Freestone *et al.* 2003, Henderson *et al.* 2005, Leslie *et al.* 2006, and papers in this volume). The use of strontium isotope ratios for investigating human skeletal material is explored for the lessons it may provide for the study of vitreous materials

The strontium isotope analysis of skeletal material was initially undertaken to gain information about diet and migration patterns (e.g. Ericson 1985). Strontium will substitute for calcium in skeletal material, including bone, horn and teeth, and largely derives from food ingested. It is widely accepted that there is negligible strontium *isotope* fractionation in the food chain (e.g. Sealy *et al.* 2002, 399) and that the strontium isotopes in such material will reflect the values in local plant material, which in turn reflects the local geological strontium isotope ratios (Evans *et al.* 2006, 312). Therefore, 'strontium isotope compositions of bones and teeth match those of the diets of the individuals, which in turn are *assumed* to reflect the strontium isotope composition of the local geology' (Price *et al.* 2002, 118, original emphasis). Much of the recent research into the strontium isotope ratios in human skeletal material has focussed on distinguishing 'local' and 'non-local' individuals, and identifying individuals who have migrated.

Some of the most dramatic applications of strontium isotope analysis on human skeletal material have targeted assemblages where there are *a priori* grounds for suspecting that some proportion of the population has migrated (e.g. Evans *et al.* 2006, Montgomery *et al.* 2003, Schweissing and Grupe 2003). Where the strontium isotope ratio in skeletal material is close to that expected for the underlying geology, it is deduced that the individual was born and lived 'locally'. Where the strontium isotope ratio differs, it is assumed that the individual migrated. In some cases, the strontium isotope ratios are dramatically different, and so distinguishing 'local' and 'non-local' is relatively straightforward (Evans *et al.* 2006, Montgomery *et al.* 2003); however, in other cases the results are usually less equivocal (e.g. Bentley *et al.* 2003, Bentley *et al.* 2004). Great potential has been shown by determining the strontium isotope ratios in teeth and bones from the same individuals. The mineral fraction of bone undergoes continuous replacement during life and the strontium isotope ratios in bone reflect diet over the last decade or so of an individual's life. The enamel of teeth does not undergo such change during life and the strontium isotope ratios in early teeth (premolars) will reflect diet during the early years. Where the strontium isotope ratios in an individual's bones and teeth differ, it is concluded that some sort of migration has occurred (Evans *et al.* 2006, Schweissing and Grupe 2003).

Strontium isotope ratios can provide insights into the diet and behaviour of individuals, but recent work is beginning to highlight some of the problems with this technique. These problems are numerous and complex, but the most pressing are the lack of unique isotope ratios and the issue of biologically available strontium. Some workers in this field have confidently described strontium isotopes as providing a 'geographical signature' (e.g. Bentley and Knipper 2005, 630) and have implied that strontium isotopes are unique. While two geologically contrasting regions may give rise to differing strontium isotope ratios, this cannot be universally assumed. The proportion of ^{87}Sr in a rock is governed by the original Rb/Sr ratio and the age of the rock, and so it is possible for different types of rocks (with different original Rb/Sr ratios) *and* of different ages to end up with similar proportions of ^{87}Sr. While there are generally few difficulties in distinguishing between the strontium isotope ratios of old rocks (e.g. Cambrian or Pre-Cambrian), and geologically recent marine sediments, other rocks are much more problematic. When using the strontium isotope ratio seawater curve (Fig. 5.1) the fact that deposits of more than one date can have the same strontium isotope ratio is often overlooked. For example, a strontium isotope ratio of 0.7080 could indicate rocks that were 29, 242, 283, 326, 334, 349, 382, 398, or 457 million years old (i.e. Oligocene, Permian, Carboniferous, Devonian, Silurian or Ordovician). Clearly a strontium isotope ratio will not provide a unique value that can be traced to a single source. Even if there were no overlaps in the strontium isotope ratio seawater curve, a ratio could indicate the age of the rocks but could not distinguish between rocks of the same age and character but in different regions. The chalks in East Yorkshire and Dorset are likely to have very similar strontium isotope ratios but are over 300 km apart. This uncertainty is further increased if one considers that ^{87}Sr will tend to be enriched in older rocks, especially those with high original Rb/Sr ratios. This is even seen in earlier published versions of the strontium isotope seawater curve (Burke *et al.* 1982) where many measured values (on carefully selected samples) have higher than expected strontium isotope ratios.

The problems with relating a strontium isotope ratio to a particular geological source are further complicated by the fact that the strontium isotope ratios in the underlying geology do not always correspond exactly with those in the biosphere.

The collection of strontium isotope data from a wide range of geological and biological samples shows that 'local strontium isotope values in the food chain cannot be taken directly from known values for bedrock geology' (Price *et al.* 2002, 119). The potential mismatch between geological substrate and the biosphere can arise from the nature of rock-weathering processes, atmospheric contributions and modern farming practices. Most of the available geological strontium isotope data are derived from whole rock samples, but the strontium concentrations and strontium isotope ratios can vary considerably between different minerals within a rock. It is not uncommon for strontium and rubidium to be concentrated in different minerals in a rock, e.g. rubidium will tend to be concentrated in some minerals (e.g. potassium feldspars), which will then have $^{87}Sr/^{86}Sr$ ratios that are higher than the whole rock values (Dickin 2005, 44, fig 3.2). Importantly, the proportion of ^{87}Sr released to the biosphere can differ from the whole rock values if the different minerals in a rock weather at different rates. The ratio of strontium isotopes that are available to biological organisms can be further altered by the influence of strontium derived from the atmosphere. Rainwater contains small amounts of strontium (e.g. Herut *et al.* 1993, Negrel and Roy 1998, Kanayama *et al.* 2002), but in some environments this atmospheric contribution may be significant. The strontium isotope ratios in rainwater derive from seawater spray, recent marine and mineral dust weathered from soils and geological deposits (Herut *et al.* 1993), which may be thousands of kilometres distant from where the rain falls (e.g. Kanayama *et al.* 2002).

A number of different avenues have been explored in attempting to resolve some of the problems outlined above. On the whole direct testing of modern ground waters has been avoided because of concerns that use of modern pesticides may significantly alter the 'natural' strontium isotope ratios (e.g. Sealy *et al.* 1991). Another approach which is being increasingly pursued is the analysis of skeletal material from archaeological herbivores, which could provide an indication of the 'local' biologically-available strontium isotope ratio (e.g. Bentley *et al.* 2003, Bentley *et al.* 2004, Montgomery *et al.* 2003, Price *et al.* 2002).

The potential for strontium isotope mismatch between geology and the overlying biosphere is well illustrated by a recent study from the United Kingdom (Evans *et al.* 2006). This study was undertaken on an assemblage where the archaeological and/or historical context provided *a priori* grounds for suspecting that at least some of the individuals were not born 'locally'. There were two burial rites employed at the Roman cemetery at Lankhills, Winchester: one which was typical of southern Roman Britain and one which was 'foreign' or 'intrusive' (Clarke 1979). The 'foreign' burial rite has been interpreted as evidence for the historically attested settlement of people from the region of modern Hungary (Clarke 1979).

The analysis of tooth enamel samples from individuals with a 'foreign' burial rite showed more variation in strontium isotope ratios than those with the 'local' burial rite. This supported the theory that those with a 'foreign' burial rite were non-indigenous but suggested that they did have a common origin. The control group (i.e. those with a local burial rite), however, raises some important points. This group included seven individuals with $^{87}Sr/^{86}Sr = 0.70849 \pm 0.00027$ (2σ), plus two outliers (whose oxygen isotopes indicated that they were not local). Evans *et al.* (2006, 270) suggest that the strontium isotope ratio in the local group is 'very close to the predicted value'; however, the late Cretaceous chalk in the Winchester region should have strontium isotope ratios around 0.7077 (McArthur *et al.* 2001). A strontium isotope ratio of 0.70849 would actually suggest a deposit only 19 million years old, that is Miocene in date. The strontium isotope ratios for the 'local' population at Winchester, therefore, could be taken to indicate that none of them were born locally! A more likely explanation is that the 'local' population obtained at least some of their strontium from rainwater which contained a higher strontium isotope ratio than the local geology. In addition, the local diet may have included a significant marine component, such as oysters (see Biddle 1975, 300, 302).

There are lessons that can be learnt from the recent use of strontium isotopes for the analysis of skeletal material: the convolutions of the seawater curve, the complex nature of most geological deposits and bio-availability. The interpretation of strontium isotope ratios with reference to the seawater curve is another area where greater caution should be exercized. The curve is not a simple one and it contains many inflections and turning points, so that rocks of many different ages can give the same isotopic ratio. There are virtually no strontium isotope ratios that can be thought of as 'unique'. The seawater curve is often treated as if it provides information on all rocks and not just marine carbonates. Rocks of the same age will not always have the same isotopic ratios depending on their composition; those with elevated levels of rubidium (e.g. feldspars and some clay minerals) will tend to have higher strontium isotope ratios than contemporary marine carbonate rocks. The strontium isotope ratios of humans, animals and plants will not always reproduce the values obtained from underlying geology. Soil formation processes, rainfall patterns and diet can result in the bio-available strontium being isotopically different from that in the underlying geology. Diet is potentially of great significance for humans, especially where marine resources are heavily used. The influence of human choices in diet and the effect that these may have on strontium isotopes have not been sufficiently explored. It is possible that different individuals within a community would have differing strontium isotopes depending on diet (e.g. varying proportions of marine foods) and that diet may be governed by age, gender and/or social status. Differences in strontium isotope ratio are clearly produced by diet, but human diets are influenced by complex human factors.

Strontium isotope analysis of glass

The use of strontium isotopes to understand the raw materials used in making glass has been a relatively recent phenomenon. Initial work by Wedepohl and Bauman (2000) using strontium and lead isotope ratios to study Roman glass in Germany has been followed by strontium, lead and oxygen isotope studies of glass from the eastern Mediterranean and the Middle East (Degryse *et al.* 2006, Freestone *et al.* 2003, Henderson *et al.* 2005, Leslie *et al.* 2006, and papers in this volume).

The initial work by Wedepohl and Bauman (2000) was undertaken on two samples of fourth-century production waste from Hambach (Eifel, Germany) and four glass vessels from nearby Roman burials. All the samples had similar strontium isotope ratios; the average for all six samples is 0.70887. Wedepohl and Baumann argued that limestones have much higher strontium isotope ratios and that they could not be the source of the calcium and strontium in the glass. They concluded that the strontium derives from the use of fresh marine shells.

Freestone *et al.* (2003) have shown that soda-rich natron and plant ash glasses from the Levant and Egypt can be distinguished using the strontium concentrations and isotope ratios (Fig. 5.2). The glasses manufactured on the Levantine coast (at Bet Eli'ezer and Bet She'an) have strontium isotope ratios that are close to the modern marine value, which is interpreted as evidence for the use of fresh marine shells (serendipitously present in the local marine sands). The plant ash glass produced at Banias (Israel) has a much lower strontium isotope ratio which is thought to derive from plant ash grown in areas with Jurassic limestone outcrops which includes Banias. The lower strontium isotope ratios of the Tel el Ashmunein (Egypt) natron glass also suggest that the strontium derives from a pre-recent geological source; in this case limestone, which is also assumed to be fortuitously present in the glassmaking sand.

Work by Leslie *et al.* (2006) confirmed that sites producing soda-rich plant ash glass often have lower strontium isotope ratios than those producing natron-based glass. They noted that the isotopic ratio ranges for Banias and Tyre (both sites producing soda-rich plant ash glass) define separate and relatively tight groups. It is suggested that these differences may reflect differences in the local geology, although the importance of bio-available strontium and its relationship with underlying geologies are acknowledged. Leslie *et al.* (2006) also caution that human behaviour (e.g. different plant ash collection strategies) could have a significant impact on the strontium isotope ratios in the finished glass.

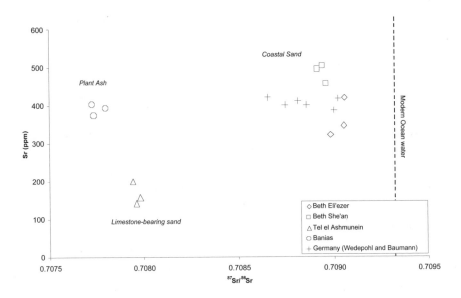

Fig. 5.2
Strontium concentrations and isotopic ratios for some Roman, Byzantine and Islamic glasses and glassworking waste (Freestone *et al.* 2003, Fig. 3)

Henderson *et al.* (2005) analysed samples of natron and plant ash glass as well as glassmaking materials (pebbles and plants) from the Islamic glass production centre of al-Raqqa (Syria). The results confirmed the work of Freestone *et al.* (2003) that plant ash glasses have lower strontium isotope ratios than natron glasses. The al-Raqqa plants have strontium isotope ratios that are close to those of the plant ash glasses, confirming that plant ashes were used and that these were probably obtained locally. The similarity between the strontium isotope ratios for the al-Raqqa natron glass samples and those from Bet Eli'ezer and Bet She'an was interpreted as evidence that the natron glass was imported from Levantine production centres.

Degryse *et al.* (2006) analysed eleven fragments of early Byzantine glass from Sagalassos (in modern Turkey). The strontium isotope ratios provided similar results to those obtained by Freestone *et al.* (2003) and indicated that some of the glass had been produced by recycling and mixing of raw glasses from two separate sources.

Where does the strontium in glass come from?

Most glasses contain small amounts of strontium, but the complex nature of raw materials used in glassmaking means that the strontium may derive from several different components. The silica source will provide varying amounts of strontium: in the case of flint or quartz pebbles almost none (Henderson *et al.* 2006, 670), but some glassmaking sands, especially those which incorporate some calcareous material, may provide substantial amounts. The calcareous material present in glassmaking sand will usually be fragments of recent seashells or of sedimentary carbonate rocks (e.g. chalk or limestone). Recent seashells will have a strontium isotope ratio similar to that of modern seawater, but sedimentary carbonate rocks could yield varying isotope ratios depending on the age and composition of the rocks. Limestones were deliberately added to glass batches, but this does not seem to have occurred before the middle of the 18th century. Glassmaking sands may also incorporate a range of minerals (e.g. feldspars) which may contain strontium. In these cases the total strontium contribution may be small but the isotope ratios could be highly variable: minerals from fairly young igneous rocks may have isotope ratios as low as 0.704, while those from old rocks may have ratios above 0.710. The contribution of even small amounts of strontium with a very high or low isotope ratio may have a dramatic effect on the isotopic ratio in the finished glass.

Strontium may also be contributed by the flux. The flux commonly employed until the 10th century or so was natron (Shortland *et al.* 2006) which is unlikely to have made a significant contribution to the total strontium content of the glass (Degryse *et al.* 2006, 497). Plant ashes used as the flux in glassmaking have made very significant contributions to the total strontium content of the glass. The amount of strontium deriving from the plant ash and its isotopic ratio will vary depending on the nature of the plants ashed and the underlying geology. Human choices could hence potentially have significant impacts on the strontium isotope ratios in glass.

Strontium in some post-medieval glass

The recent excavation of the 17th century glass production site at Silkstone, England (Fig. 5.3) included the chemical analysis of over 400 samples of glass and glassworking debris carefully selected from well stratified and dated contexts (Dungworth and Cromwell 2006). For over fifty years this glasshouse produced two types of glass: a pale green or colourless glass (probably used for tablewares) and a dark green glass (used for wine bottles and some tablewares). In the earliest phase of production (*c*. 1660 to *c*. 1670) the pale glass was a mixed alkali glass with a high strontium content. On the basis of strontium: calcium ratios it was suggested that the strontium derived from the use of seaweed ash as a flux (Dungworth and Cromwell 2006, 174). The proposition that seaweed ash was used as a flux in post-medieval glassworking is supported by 18th century documentary references which mention kelp (seaweed ash) as part of the stock of a number of glasshouses (e.g. Berg and Berg 2001, Buckley 2003, Cable 2001). While most post-medieval glass contains less than 0.1wt% SrO, a few other post-medieval glassworking sites have also begun to yield samples of glass containing up to 0.46wt% SrO (e.g. Jackson 2005, 124). The suggestion that seaweed was used in the manufacture of some post-medieval glass has been tested by determining strontium isotope ratios. Any glass made using seaweed ash is likely to obtain almost all of its strontium from the seaweed ash and so will have an isotope ratio close to that of modern seawater (0.7092).

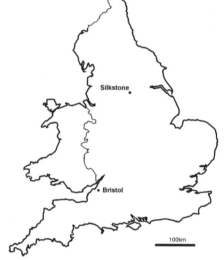

Fig. 5.3
Map of England and Wales showing the locations of the glasshouses at Silkstone and Bristol

Samples and Methods

Twelve samples were analysed to determine the $^{87}Sr/^{86}Sr$ ratios: one sample of modern seaweed and eleven samples of glassworking debris. The eleven samples of glassworking debris include:

- three samples of the Silkstone phase 1 mixed alkali glass (average SrO = 0.27 wt%) which are likely to have been made using seaweed (samples 5–7)
- three samples of the Silkstone phase 2 mixed alkali glass (average SrO = 0.05 wt%) which were made using an unidentified (terrestrial) plant ash (samples 8–10)

- three samples of mixed alkali glass from Bristol (St Thomas Street, Jackson 2004, average SrO = 0.41 wt%) which are likely to have been made using seaweed (samples 2–4)
- two samples of the Silkstone phase 1–2 dark green high-lime low-alkali glass (average SrO = 0.06 wt%) which were made using an unidentified (terrestrial) plant ash (samples 11 and 12)

The selection of samples sought to determine whether seaweed had been used to manufacture the high strontium glasses and to compare these results with those from low strontium glasses.

The chemical compositions of the samples were determined using an energy dispersive X-ray fluorescence (EDXRF) spectrometer and an energy dispersive X-ray spectrometer attached to a scanning electron microscope (SEM-EDS). The EDXRF spectrometer (an EDAX Eagle, operated at 40kV and 0.1mA for 100 seconds livetime) provided low detection limits and good precision for most of the heavier elements (Fe–Pb). The SEM-EDS (an Oxford Instruments germanium detector with 25kv accelerating voltage and 1.5nA current for 100 seconds livetime) provided results for the lighter elements (Na–Fe). Both instruments were calibrated using a range of suitable reference materials (eg NIST and Corning), although the readily available CRMs contain lower concentrations of strontium than some of the glass samples. In all cases the oxide weight percentages were calculated stoichiometrically, but at least some of the cations present in the seaweed ash are likely to be present as carbonates or chlorides. The results are given in Table 5.2.

For isotope analysis, samples (several hundred milligrams) were weighed into Teflon screw-top beakers and dissolved in a 3:1 mixture of 22 N HF and 14 N HNO_3 on a hot plate. Solutions were dried and dissolved in aqua regia. Strontium was chemically separated and purified by conventional anion exchange methods on 250 μl columns containing EICHROM Sr resin using the methods of Horwitz *et al.* (1991 a and b). Measurements were performed on a six-collector FINNIGAN MAT 261 thermal ionisation mass spectrometer (TIMS) running in static multi-collection mode. Sr isotopic ratios were normalized to $^{88}Sr/^{86}Sr = 0.1194$. Repeated static measurements of the NBS 987 standard over the duration of the study yielded an average $^{87}Sr/^{86}Sr$ ratio of 0.71025 ± 4 (2σ, n=22). Total procedural blanks (n=6) did not exceed 30 pg Sr.

#	Material	Provenance	$^{87}Sr/^{86}Sr$	Na_2O	MgO	Al_2O_3	SiO_2	P_2O_5	SO_3	Cl	K_2O	CaO	TiO_2	MnO	Fe_2O_3	SrO	PbO
1	Seaweed	Portsmouth	0.70932	8.7	8.2	0.6	1.7	1.7	33.3	0.6	14.9	25.6	0.11	2.63	1.34	0.74	<0.05
2	MA glass	St Thomas St, Bristol	0.70941	6.9	4.7	3.4	68.4	0.9	0.2	0.6	4.9	9.1	0.10	0.04	0.56	0.38	<0.05
3	MA glass	St Thomas St, Bristol	0.70931	6.7	5.2	3.7	65.5	1.0	0.3	0.6	4.6	11.3	0.11	0.04	0.74	0.42	<0.05
4	MA glass	St Thomas St, Bristol	0.70919	7.8	4.8	3.4	67.0	0.9	0.3	0.7	3.8	10.0	0.13	0.05	0.71	0.38	<0.05
5	MA glass	Silkstone, Yorkshire	0.70930	8.2	4.7	3.2	62.4	1.2	0.2	0.7	5.9	8.8	0.19	0.45	1.08	0.25	2.96
6	MA glass	Silkstone, Yorkshire	0.70934	8.1	4.9	2.9	62.6	1.5	0.2	0.7	5.7	9.0	0.23	0.46	1.06	0.26	2.58
7	MA glass	Silkstone, Yorkshire	0.70931	8.0	4.8	3.0	62.5	1.3	0.2	0.8	5.9	9.1	0.23	0.53	1.17	0.28	2.57
8	MA glass	Silkstone, Yorkshire	0.70901	7.1	2.6	0.9	68.8	0.2	0.2	0.6	6.9	10.1	0.15	1.12	0.76	0.05	0.15
9	MA glass	Silkstone, Yorkshire	0.70888	6.6	2.7	1.1	69.7	0.2	0.2	0.4	6.7	10.1	0.10	1.10	0.99	0.05	0.21
10	MA glass	Silkstone, Yorkshire	0.70901	6.6	2.8	1.2	69.4	0.2	0.2	0.6	6.9	9.9	0.18	1.12	0.50	0.05	0.21
11	HLLA glass	Silkstone, Yorkshire	0.71443	1.7	4.0	4.6	55.5	3.0	0.4	0.3	9.9	17.3	0.22	0.62	2.42	0.06	0.11
12	HLLA glass	Silkstone, Yorkshire	0.71791	1.7	4.5	4.0	55.2	2.1	0.3	0.4	8.4	20.2	0.28	0.44	2.53	0.04	<0.05

Table 5.2
Composition of seaweed and glass samples studied in Dungworth *et al.* (this volume). The composition of the seaweed was determined by EDXRF. The composition of the glasses was determined by a combination of SEM-EDS and EDXRF (the latter for MnO, Fe_2O_3, SrO and PbO).

Results

The sample of modern seaweed (collected from the beach at Portsmouth, England) gave a strontium isotope ratio of 0.70932, which is comparable with measurements of modern seawater and marine shells (Burke *et al.* 1982). The samples of high-strontium mixed-alkali glass from Silkstone and Bristol also gave isotope ratios that clustered around the same value (0.70931±0.00014 at 2δ), which is compatible with these glasses getting almost all of their strontium from a fresh marine source (Fig. 5.4). Taking this in combination with the high strontium concentration of these glasses, it seems most likely that they were made using seaweed ash. The Bristol samples of mixed-alkali (seaweed) glass show the highest degree of variability which is possibly due to the contribution of strontium from other batch ingredients. These analyses have confirmed the use of seaweed ash for glassmaking in the 17th century. While the use of seaweed ash is unsurprising at Bristol, which had a port and easy communication with the coasts of western Britain, it is striking that the glassmakers at Silkstone obtained at least some of their raw materials from areas at least 120km away.

The three samples of low-strontium mixed-alkali glass from Silkstone (Phase 2) have lower strontium isotope ratios (0.70907 ± 0.00044 at 2σ). The strontium concentrations and isotopic ratios of these samples (Fig. 5.4) are comparable with those of the Roman and Byzantine natron glasses analysed by Wedepohl and Baumann (2000) and Freestone *et al*. (2003); however, the bulk compositions are significantly different. The Phase 2 Silkstone mixed-alkali glass has a bulk composition which is broadly similar to that of Phase 1 (the seaweed mixed-alkali glass). It is likely that the flux employed to make the Phase 2 mixed-alkali glass was made using a soda-rich terrestrial plant ash, but it is far from clear which plant(s) were used. Christopher Merrett, writing in the 17th century, suggests that a range of plants grown in England could be used to produce the finer qualities of glass, including Saltwort, Glasswort and Seagrass (Cable 2001). Assuming that the strontium isotope ratios of the Phase 2 Silkstone mixed-alkali glass reflect the geology on which the plants grew, these plants are likely to have grown in areas with mid-Miocene to early Pleistocene deposits. There are few such deposits close to Silkstone, but they are common round the coasts of England (especially the south and east).

The last two samples analysed are dark green high-lime low-alkali glasses from Silkstone which proved to have extremely high strontium isotope ratios (Fig. 5.5), which suggests an association with relatively old rocks in which ^{87}Sr would be enriched by the decay of ^{87}Rb. The nearest areas with suitable rocks are the far North-West and South-West of England, Scotland and Wales which are at least 130km from Silkstone.

The strontium isotope results for Silkstone provide an insight into the procurement of glassmaking raw materials in the post-medieval period (Crossley 1998). The largely unstated assumption when using scientific methods to understand the raw materials used in glassmaking is that they would have been locally obtained. At Silkstone, however, it appears that three different fluxes were obtained from sources over 100km away. This is despite the likelihood that the products of the Silkstone glasshouse had no more than a regional distribution.

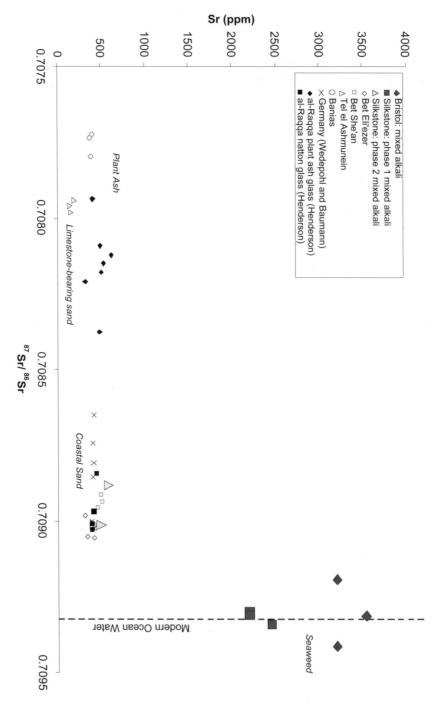

Fig. 5.4
Strontium concentrations and isotopic ratios for some post-medieval glassworking waste

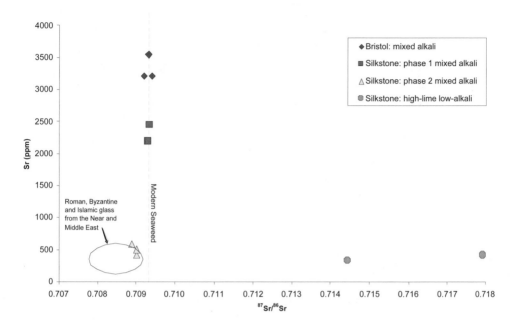

Fig. 5.5
Strontium concentrations and isotopic ratios for some post-medieval glassworking waste
(detail)

Conclusion

The use of strontium isotope analysis for characterising vitreous materials is a very
recent development, but results have shown similarities and differences between
different glassworking sites. The results have also been interpreted in terms of the
types of raw materials used in their manufacture. The use of strontium isotopes
in understanding the movement of individuals in the past has been established
for somewhat longer and offers several significant lessons. While the number of
analysed samples of skeletal material has grown, there are still few analyses of
biologically available strontium with which these can be compared. The inference
of strontium isotopes from the strontium seawater curve and geological maps is
not particularly rigorous. Where strontium isotope ratios differ it can be concluded
that the two groups have different sources, but identifying a 'strontium isotope
signature' and a unique source is not really possible. The variability of geochemical
processes is likely to mean that there will always be multiple potential sources,
although some of these may be discounted using other types of evidence (e.g.
other isotopes and elemental concentrations).

The analysis of a sample of seaweed and six samples of high-strontium mixed alkali glass showed that they all shared the same average isotopic ratio and confirmed that the glasses were made using seaweed ash. The remaining glass samples have low concentrations of strontium and isotopic ratios which differ significantly from the modern seawater value. It is not possible to provide exact geographical origins for the various raw materials employed at Silkstone; however, the three glass types examined here all appear to have been made using at least some raw materials which were obtained over a surprisingly wide area.

References

T. Berg, P. Berg (eds.) 2001, R R Angerstein's Illustrated Travel Diary, 1753–1755, London Science Museum.

R.A. Bentley, C. Knipper, 2005, Geographical patterns in biologically available strontium, carbon, and oxygen isotope signatures in prehistoric SW Germany, Archaeometry, 47, 629–644.

R.A. Bentley, R. Krause, T.D. Price, B. Kaufmann, 2003, Human mobility at the early Neolithic settlement of Vaihingen, Germany: evidence from strontium isotope analysis, Archaeometry, 45, 471–486.

R.A. Bentley, T.D. Price, E. Stephan, 2004, Determining the 'local' 87Sr/ 86Sr range for archaeological skeletons: a case study from Neolithic Europe, Journal of Archaeological Science, 31, 365–375.

M. Biddle, 1975, Excavations at Winchester, 1971. Tenth and final interim report: Part II, Antiquaries Journal, 55, 295–337.

P. Budd, A. Millard, C. Chenery, S. Lucy, C. Roberts, 2004, Investigating population movement by stable isotope analysis: a report from Britain, Antiquity, 78, 127–141.

F. Buckley, 2003, Old English Glass Houses, Society of Glass Technology.

W.H. Burke, R.E. Denison, E.A. Hetherington, R.B. Koepnick, H.F. Nelson, J.B. Otto, 1982, Variation of seawater $^{87}Sr/^{86}Sr$ throughout Phanerozoic time, Geology, 10, 516–519.

M. Cable (ed.) 2001, The World's Most Famous Book on Glassmaking. The Art of Glass, by Antonio Neri, translated into English by Christopher Merrett, Society of Glass Technology.

G. Clarke, 1979, Pre-Roman and Roman Winchester. Part II: the Roman cemetery at Lankhills, University of Oxford.

D.W. Crossley, 1998, The English glassmaker and his search for raw materials in the 16th and 17th centuries, in: P. McCray (ed.) The Prehistory and History of Glassmaking Technology. Westerville, American Ceramic Society, 167–179.

A.P. Dickin, 2005, Radiogenic Isotope Geology, 2nd edition, Cambridge University Press.

P. Degryse, J. Schneider, U. Haack, V. Lauwers, J. Poblome, M. Waelkens, Ph. Muchez, 2006, Evidence for glass 'recycling' using Pb and Sr isotopic ratios and Sr-mixing lines: the case of early Byzantine Sagalassos, Journal of Archaeological Science, 33, 494-501.

D. Dungworth, T. Cromwell, 2006, Glass and pottery manufacture at Silkstone, Yorkshire, Post-Medieval Archaeology, 41, 161–190.

J.R. Ericson, 1985, Strontium isotope characterization in the study of prehistoric human ecology, Journal of Human Evolution, 14, 503–514.

J.A. Evans, C.A. Chenery, A.P. Fitzpatrick, 2006, Bronze Age childhood migration of individuals near Stonehenge, revealed by strontium and oxygen isotope tooth enamel analysis, Archaeometry, 48, 309–321.

J.A. Evans, N. Stoodley, C. Chenery, 2006, A strontium and oxygen isotope assessment of a possible fourth century immigrant population in a Hampshire cemetery, southern England, Journal of Archaeological Science, 33, 265–272.

G. Faure, T.M. Mensing, 2005, Isotopes. Principles and applications, Wiley.

G. Faure, J.L. Powell, 1972, Strontium Isotope Geology, Springer-Verlag.

I.C. Freestone, K. A. Leslie, M. Thirlwall, Y. Gorin-Rosen, 2003, Strontium isotopes in the investigation of early glass production: Byzantine and early Islamic glass from the Near East, Archaeometry, 45, 19-32.

G. Grupe, T.D. Price, P. Schröter, F. Söllner, C.M. Johnson, B.L. Beard, 1997, Mobility of Bell Beaker people revealed by strontium isotope ratios of tooth and bone: a study of southern Bavarian skeletal remains, Applied Geochemistry, 12, 517–255.

J. Henderson, 1989, The scientific analysis of ancient glass and its archaeological interpretation, in: J. Henderson (ed.), Scientific Analysis in Archaeology, Oxford University Committee for Archaeology, 30–62.

J. Henderson, J.A. Evans, H.J. Sloane, M.J. Leng, C. Doherty, 2005, The use of oxygen, strontium and lead isotopes to provenance ancient glasses in the Middle East, Journal of Archaeological Science, 32, 665-673.

B. Herut, A. Starinsky, A. Katz, 1993, Strontium in rainwater from Israel: Sources, isotopes and chemistry, Earth and Planetary Science Letters, 120, 77–84.

D.A. Hodell, R.L. Quinn, M. Brenner, G. Kamenov, 2004, Spatial variation of strontium isotopes ($^{87}Sr/^{86}Sr$) in the Maya region: a tool for tracking ancient human migration, Journal of Archaeological Science, 31, 585–601.

E.P. Horwitz, M.L. Dietz, D.E. Fischer, 1991a, SREX: a new process for the extraction and recovery of strontium from acidic nuclear waste streams, Solvent Extraction and Ion Exchange, 9, 1–25.

E.P. Horwitz, M.L. Dietz, D.E. Fischer, 1991b, Separation and preconcentration of Sr from biological, environmental and nuclear waste samples by extraction chromatography using a crown ether, Analytical Chemistry, 63, 522–525.

R. Jackson, 2004, Archaeological excavations at Nos. 30–38 St Thomas Street & No. 60 Redcliff Street, Bristol, 2000, Bristol & Avon Archaeology, 19, 1–63.

R. Jackson, 2005, Excavations on the site of Sir Abraham Elton's glassworks, Cheese Lane, Bristol, Post-Medieval Archaeology, 39, 92–132.

S. Kanayama, S. Yabuki, F. Yanagisawa, R. Motoyama, 2002, The chemical and strontium isotope composition of atmospheric aerosols over Japan: the contribution of long-range-transported Asian dust (Kosa), Atmospheric Environment, 36, 5159–5175.

C. Kendall, M.G. Sklash, T.D. Bullen, 1995, Isotope tracers of water and solute sources in catchments, in: S.T. Trudgill (ed.) Solute Modelling in Catchment Systems, Wiley, 261–303.

K.A. Leslie, I.C. Freestone, D. Lowry, M. Thirlwall, 2006, Provenance and technology of near Eastern glass: oxygen isotopes by laser fluorination as a compliment to Sr, Archaeometry, 48, 253-270.

J.M. McArthur, R.J. Howarth, T.R. Bailey, 2001, Strontium isotope stratigraphy: LOWESS version 3: best fit to the marine Sr-isotope curve for 0–509 Ma and accompanying look-up table for deriving numerical age, Journal of Geology, 109, 155–170.

J. Montgomery, J. Evans, T. Neighbour, 2003, Sr isotope evidence for population movement within the Hebridean Norse community of NW Scotland, Journal of the Geological Society, London, 160, 649–653.

S. Moorbath, J.D. Bell, 1965, Strontium isotope abundance studies and Rubidium—Strontium age determinations on Tertiary igneous rocks from the Isle of Skye North-West Scotland, Journal of Petrology, 6, 37–66.

P. Negrel, S. Roy, 1998, Chemistry of rainwater in the Massif Central (France): a strontium isotope and major element study, Applied Geochemistry, 13, 941–952.

T.D. Price, J.H. Burton, R.A. Bentley, 2002, The characterization of biologically available strontium isotope ratios for the study of prehistoric migration, Archaeometry, 44, 117–135.

T.D. Price, C.M. Johnson, J.A. Ezzo, J. Ericson, J.H. Burton, 1994, Residential-mobility in the prehistoric Southwest United States—a preliminary study using strontium isotope analysis, Journal of Archaeological Science, 21, 315–330.

T.D. Price, C. Knipper, G. Grupe, V. Smrcka, 2004, Strontium isotopes and prehistoric human migration: the Bell Beaker Period in central Europe, European Journal of Archaeology, 7, 9–40.

M.M. Schweissing, G. Grupe, 2003, Stable strontium isotopes in human teeth and bone: a key to migration events of the late Roman period in Bavaria, Journal of Archaeological Science, 30, 1373–1383

J.C. Sealy, N.J. Vandermerwe, A. Sillen, F.J. Krueger, H.W. Krueger, 1991, Sr-87 Sr-86 as a dietary indicator in modern and archaeological bone, Journal of Archaeological Science, 18, 399–416

A.J. Shortland, L. Schachner, I.C. Freestone, M. Tite, 2006, Natron as a flux in the early vitreous materials industry: sources, beginnings and reasons for decline, Journal of Archaeological Science, 33, 521–530.

Z.A. Stos-Gale, 1989, Lead isotope studies of metal and the metal trade in the Bronze Age Mediterranean, in: J. Henderson (ed.) Scientific Analysis in Archaeology, Oxford University Committee for Archaeology, 274–301.

K.H. Wedepohl, A. Baumann, 2000, The use of marine molluscan shells in the Roman glass and local raw glass production in the Eifel area (Western Germany), Naturwissenschaften, 87, 129-132.

Medieval and postmedieval Hispano-Moresque glazed ceramics: new possibilities of characterization by means of lead isotope ratio determination by Quadrupole ICP-MS

Paz Marzo, Francisco Laborda, Josefina Pérez-Arantegui

Introduction

Lead has been used on glazed ceramics since the 10th century BC (Cooper 1998) because of the technical advantages that it presents, such as less risk of glaze crazing, lower surface tension, higher refractive index and greater brilliance of the surfaces of glazed ceramics (Tite *et al*. 1998). The amount of lead used has not been constant at all times, but its use has been continuous since the Roman era in the Mediterranean area and the near East. In the Iberian Peninsula, glazed ceramics are already found in archaeological sites of the Roman period, especially as imported objects or sometimes as possible local products. This is the case of ceramics decorated with high-lead glazes, in brown, green or yellow, with a high lead oxide content in the glaze (50-65 wt% PbO) (Beltrán 1990, Casas and Merino 1990, López 1978, Pérez-Arantegui *et al*. 1995). However, it was probably after the expansion of the Islamic culture from the 8th century AD that this type of glaze was widely manufactured in the Iberian Peninsula. The lead-glaze tradition continued with the production of tin-opacified glazes from the 10th century AD onwards, always as lead-rich glazes, and with extensive use in Hispano-Moresque ceramics during the medieval and post-medieval periods.

All analytical results published show that glazes made in the Iberian Peninsula are always lead glazes (between 25 and 55 wt% PbO) with low proportions of alkaline elements (less than 5 wt% Na_2O+K_2O), either transparent or tin-opacified. In Table 6.1 we can see some examples of both types of glazes and different decorations, with characteristic proportions of lead oxide and low alkaline content (Molera *et al*. 1996, Molera *et al*. 2001, Pérez-Arantegui and Lapuente 2003, Pérez-Arantegui *et al*. in press).

One of the main aims in ancient ceramic research is material characterization for provenance and technological study. In the case of glazed ceramics, the knowledge of the raw materials used to produce the glaze helps to establish better a pattern of manufacture. As a major component of the glaze, lead was an ingredient linked to significant trade contacts, and it is important to distinguish different lead sources used in diverse historical sites or periods. Because lead isotope ratios are related to mineral sources, lead isotopes analyses have been shown to give important information on provenance for several archaeological materials, especially for metals or alloys (Gale and Stos-Gale 2000).

Site	Decoration	Al_2O_3	SiO_2	K_2O	CaO	SnO_2	PbO
Zaragoza	Green-Black	0.61	46.1	3.74	3.73	5.87	38.0
Paterna*	Blue-Lustre	2.77	51.8	6.49	2.33	2.06	33.8
Paterna*	Green- Black	3.66	45.8	2.42	2.24	1.12	40.3
Teruel	Green- Black	4.11	41.7	1.61	0.63	8.34	43.2
Teruel	Blue	4.50	47.4	4.63	1.63	4.88	36.4
Muel	Lustre	3.09	49.1	4.89	2.36	5.76	34.2
Teruel	Honey	5.50	41.8	1.30	1.44	-	45.5

Table 6.1
Chemical composition (in wt%) of some ceramic glazes produced in the Iberian Peninsula (tin-opacified glazes in all but the last example), (*) Molera *et al.* (1996)

Natural lead consists of four isotopes: [208]Pb, [207]Pb, [206]Pb and [204]Pb. The first three derive partly from the radioactive decay of long-lived naturally radioactive isotopes of U and Th, but [204]Pb is non-radiogenic in origin. This is why natural lead varies in isotopic composition. As a result of the evolution of lead isotopic rock compositions and ore formation processes, the isotopic composition of lead in an ore deposit varies from deposit to deposit depending on their geological ages.

Lead isotope analysis is routinely performed by thermal ionization mass spectrometry (TIMS) with multicollector detection (Pomiès *et al.* 1998, Santos *et al.* 2004, Stos-Gale *et al.* 1995, Wolf *et al.* 2003, Yener *et al.* 1991). However, apart from the inefficient ionization yields of elements that have a high ionization potential, some major disadvantages of TIMS are the time-consuming sample preparation that is required in order to separate the element of interest from the matrix, and the relatively long time needed for data acquisition. Inductively-coupled-plasma quadrupole-mass-spectrometry (ICP-QMS) is a versatile mass spectrometric technique with higher sensitivity, simpler sample preparation procedures and higher sample throughput than TIMS. Precision of the isotope ratio measurements is the only limitation of ICP-QMS, although a relative standard deviation of around 0.1% can be achieved if sources of noise are reduced and acquisition parameters of ICP-MS are optimized. A major advantage of the plasma excitation is its very high temperature, which renders the sample composition almost irrelevant. Unlike in the use of TIMS, the element of interest does not need to be separated from the sample matrix, which results in considerably simplified and time-saving sample preparation for many applications.

Lead isotope studies carried out on glazes (Habitcht *et al.* 2000, Wolf *et al.* 2003) have allowed different lead isotope ratios to be established. The comparison of those ratios with the database ratios from lead ores points out possible sources of lead used in the manufacture of glazed pottery and suggests different patterns of resources utilized, depending on the periods and location of the workshops.

The study presented here is focused on two different aims: first, an evaluation of the performance of Inductively-Coupled-Plasma Quadrupole-Mass-Spectrometry (ICP-QMS) for lead isotope ratio measurements in the characterization of lead-glazed ceramics; secondly, the determination of lead isotope ratios in a selection of glazes, to distinguish production areas and periods. Specifically, glazed ceramics produced in the Iberian Peninsula in different periods, especially in the area of Aragon, were chosen.

Experimental

MATERIALS AND METHODS

Lead isotope ratios were measured using a Perkin Elmer SCIEX Elan 6000 ICP-QMS, equipped with a cross-flow nebulizer. The masses measured were 204, 206, 207, 208, and 202 to correct ^{204}Hg interferences.

Stocks solutions from 10 ng g^{-1} to 100 ng g^{-1} of Pb of known isotopic composition (^{204}Pb/^{206}Pb= 0.059042 ± 0.000037; ^{207}Pb/^{206}Pb= 0.91464 ± 0.00033; ^{208}Pb/^{206}Pb= 2.1681 ± 0.0008) were prepared from an isotopic standard reference material, SRM-981 (NIST, Gaitherburg, USA), using nitric acid (2% (w/w)).

SAMPLE PREPARATION

Glazes were sampled by cutting a small surface portion of about 1 cm^2. For the preparation of the sample solution, the glaze was put in 2 mL of acetic acid 4% v/v in a closed PTFE beaker at room temperature for its lixiviation during 24 hours. Finally, samples were diluted with 2% w/w HNO$_3$ to keep the range of 50 - 100 ng ml^{-1} Pb required to quantify the lead isotope ratios by ICP-QMS. Three replicates of each solution were done, with a total analysis time of less than six minutes per sample, and a standard deviation of around 0.1%, depending on the lead isotope.

Results and discussion

TIMS is a very precise and accurate technique for isotope ratio determination, although separation of the element of interest from the sample matrix is required and a relatively long time is needed for data acquisition (from one up to several hours). The ICP-QMS method proposed is simpler; lead is lixiviated from the sample just by acetic acid, and the acquisition time is reduced to less than six minutes per sample. The attainable precision (around 0.1%) is sufficient for the purpose of the glaze characterization.

Sample	Site	Glaze decoration	Date
1	Utebo tower	White tin-glaze	15th c.
2	Utebo tower	Green tin-glaze	15th c.
3	Albarracin	Honey glaze	11th c.
4	Albarracin	Green & honey glaze	11th c.
5	Zaragoza	Honey glaze	11th c.
6	Zaragoza	Green & black on white tin-glaze	11th c.
7	Muel-Zaragoza	Blue on white tin-glaze	16th c.
8	Muel-Zaragoza	Lustre	16th c.
9	Muel-Zaragoza	Lustre	16th c.
10	Teruel	Green glaze	14th c.
11	Teruel	Green & black on white tin-glaze	14th c.
12	Teruel	Blue on white tin-glaze	15th c.
13	Pechina (Almeria)	Green & black on white tin-glaze	11th c.
14	Raqqa (Syria)	Green glaze	12th c.
15	Albarracin	Green & black on white tin-glaze	11th c.
16	Albarracin	Green & black on white tin-glaze	11th c.
17	Albarracin	Green & black on white tin-glaze	11th c.
18	Albarracin	Green & black on white tin-glaze	11th c.
19	Albarracin	Honey glaze	11th c.
20	Albarracin	Honey glaze	11th c.
21	Albarracin	Honey glaze	11th c.
22	Albarracin	Honey glaze	11th c.

Table 6.2
Ceramic glaze samples from different sites and periods, chosen for the lead isotope determination

For lead isotope ratio analysis a group of glazed ceramics, described in Table 6.2 and Fig. 6.1, was selected. The glazes belonged to ceramic fragments already studied in depth in previous researches, and with very well-known compositions and microstructures (Molera *et al.* 2001, Pérez-Arantegui and Lapuente 2003,

Pérez-Arantegui *et al*. in press). Some of them were Islamic ceramics, dated to the 11th century AD, with different decorations (samples 3-6, 13, 15-22). They were produced in Zaragoza, Albarracin and Almeria, three different Taifa kingdoms of the Iberian Peninsula in that period. Other ceramics (samples 7-12) corresponded to Hispano-Moresque glazes and decorations, manufactured between the 14th and the 16th centuries AD in two different places, famous because of their ceramic production, situated in the kingdom of Aragon : Muel (labelled as Zaragoza-Mudejar), a village very close to Zaragoza, and Teruel. Finally, two other samples were analysed (samples 1, 2): two glazes from ceramic tiles used as architectural decoration dating from the 15th century AD. These tiles decorated a tower in Utebo (Zaragoza), and were a type of decoration very usual in that period in Aragon. Moreover another sample (number 14) was included specifically to test the analytical method; it corresponds to an Islamic ceramic fragment from Raqqa (Syria).

Fig. 6.1
Map showing the studied sites and some important towns in the Iberian Peninsula, including the main lead-ores areas

The results of the lead isotope ratio determination are summarized in Table 6.3. For the graphical presentation of the lead isotope data we are using the two bivariant diagrams, including the four isotopes ($^{208}Pb/^{206}Pb$ *vs*. $^{207}Pb/^{206}Pb$ and $^{206}Pb/^{204}Pb$ *vs*. $^{207}Pb/^{206}Pb$), which reflect both the time of formation of the lead ore deposit and the isotopic composition of the source region prior to formation (Fig. 6.2 and 6.3). The sample from Raqqa (Syria), included specifically for testing the analytical method, shows completely different lead isotope ratios from the Iberian samples, much lower ratios like other results in glasses from the same region (Henderson *et al*. 2005); this is why it was not included in the graphs.

Sample	^{206}Pb/^{204}Pb ± sd	^{207}Pb/^{204}Pb ± sd	^{208}Pb/^{204}Pb ± sd	^{207}Pb/^{206}Pb ± sd	^{208}Pb/^{206}Pb ± sd
1	18.258 ± 0.008	15.533 ± 0.032	37.857 ± 0.114	0.8501 ± 0.0015	2.0797 ± 0.0018
2	18.346 ± 0.017	15.695 ± 0.022	38.342 ± 0.001	0.8495 ± 0.0001	2.0900 ± 0.0028
3	18.235 ± 0.024	15.610 ± 0.014	38.569 ± 0.011	0.8556 ± 0.0005	2.1133 ± 0.0019
4	18.220 ± 0.013	15.624 ± 0.036	38.316 ± 0.013	0.8586 ± 0.0004	2.1043 ± 0.0008
5	18.184 ± 0.049	15.619 ± 0.044	38.068 ± 0.110	0.8589 ± 0.0001	2.0935 ± 0.0028
6	18.343 ± 0.051	15.660 ± 0.043	38.534 ± 0.098	0.8539 ± 0.0001	2.1041 ± 0.0020
7	18.368 ± 0.055	15.689 ± 0.061	38.953 ± 0.036	0.8541 ± 0.0007	2.1054 ± 0.0022
8	18.295 ± 0.026	15.659 ± 0.013	38.246 ± 0.084	0.8528 ± 0.0010	2.0906 ± 0.0004
9	18.377 ± 0.038	15.693 ± 0.052	38.640 ± 0.120	0.8542 ± 0.0011	2.1044 ± 0.0016
10	18.368 ± 0.022	15.744 ± 0.024	38.979 ± 0.081	0.8573 ± 0.0003	2.1223 ± 0.0021
11	18.274 ± 0.049	15.677 ± 0.043	38.588 ± 0.089	0.8579 ± 0.0001	2.1117 ± 0.0009
12	18.190 ± 0.069	15.604 ± 0.026	38.075 ± 0.111	0.8578 ± 0.0019	2.0932 ± 0.0019
13	18.344 ± 0.053	15.553 ± 0.043	38.073 ± 0.115	0.8479 ± 0.0012	2.0755 ± 0.0046
14	18.979 ± 0.058	15.705 ± 0.063	39.597 ± 0.097	0.8277 ± 0.0008	2.0863 ± 0.0026
15	18.504 ± 0.045	15.652 ± 0.029	38.672 ± 0.044	0.8457 ± 0.0001	2.0881 ± 0.0019
16	18.523 ± 0.010	15.662 ± 0.004	38.570 ± 0.042	0.8454 ± 0.0002	2.0820 ± 0.0017
17	18.556 ± 0.032	15.676 ± 0.035	38.588 ± 0.036	0.8458 ± 0.0001	2.0877 ± 0.0026
18	18.481 ± 0.053	15.666 ± 0.030	38.525 ± 0.118	0.8471 ± 0.0009	2.0827 ± 0.0016
19	18.369 ± 0.023	15.715 ± 0.032	38.889 ± 0.029	0.8564 ± 0.0009	2.1209 ± 0.0023
20	18.310 ± 0.003	15.661 ± 0.014	38.839 ± 0.013	0.8567 ± 0.0009	2.1263 ± 0.0010
21	18.328 ± 0.020	15.712 ± 0.014	38.862 ± 0.060	0.8578 ± 0.0003	2.1197 ± 0.0019
22	18.250 ± 0.023	15.627 ± 0.012	38.679 ± 0.064	0.8560 ± 0.0006	2.1230 ± 0.0009

Table 6.3
Lead isotope data for different glazed ceramic samples (mean value and standard deviation, n=3)

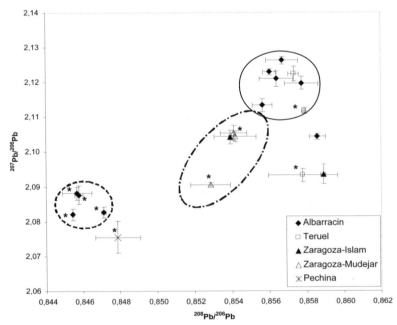

Fig. 6.2
Lead isotope ratios for glazed pottery samples (^{208}Pb/^{206}Pb vs. ^{207}Pb/^{206}Pb)

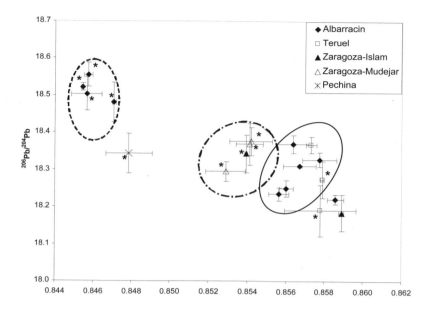

Fig. 6.3
Lead isotope ratios for glazed pottery samples (^{206}Pb/^{204}Pb vs. ^{207}Pb/^{206}Pb)

Coming back to the Iberian glazes, we can try to observe or establish some differences or similarities. By a statistical treatment of distances, we can divide the samples into three main subgroups (Fig. 6.4). The first corresponds in particular to samples of transparent high-lead glazes from Islamic Albarracin (numbers 3, 19, 20, 21 and 22). The second group includes samples of tin-glazes from Zaragoza (Islamic and Hispano-Moresque) (numbers 6, 7 and 9). Finally the third group comprises samples of Islamic tin glazes from Albarracin (numbers 15, 16, 17 and 18).

Looking at the results corresponding to Islamic glazes, the first clear difference is the division of the Albarracin data into two groups: one corresponding to the tin glazes and a second one with the transparent high-lead glazes (Fig. 6.2 and 6.3). Both groups also differ from glazes of Islamic Zaragoza and Pechina (Almeria), in the south. The samples from the three Islamic sites (Albarracin, Zaragoza and Almeria) can be distinguished and separated. The results from Hispano-Moresque glazes (Zaragoza-Mudejar and Teruel) also show different lead isotope ratios for each group, but, at the same time, are not very well separated from the (geographically close) Islamic samples, Islamic Zaragoza and Albarracin, respectively. Another interesting group of samples is the tin-glazed ceramics (labelled with an asterisk, Fig. 6.2 and 6.3) for both periods. In this case, we can observe in particular the samples produced in Albarracin and Zaragoza, with very similar lead isotope ratios inside each group, some of them very close to one another.

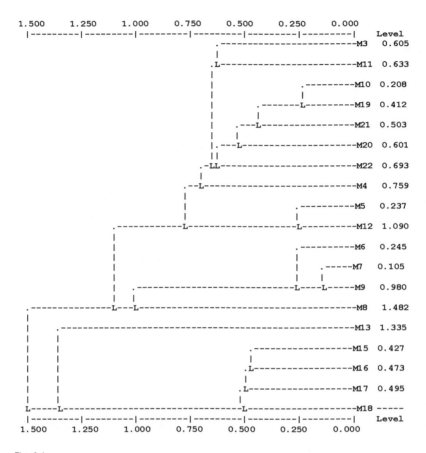

Fig. 6.4
Hierarchical clustering analysis of lead isotope ratio data

The last samples considered (samples 1 and 2), the glazes on the tiles, seem to be different from the glazes on ceramic tableware. Conclusions drawn on these two samples are less certain, however, because the tower dates to the Mudejar period (15th century AD) but the structure has suffered numerous reconstructions and restorations.

These first results on lead isotope ratios of glazes produced in the medieval and post-medieval period in the Iberian Peninsula show the possibility of establishing a manufacture pattern, in different areas and periods. On the one hand, Islamic samples, including Hispano-Moresque glazes, produced in different areas, even relatively close together as in the case of Zaragoza and Albarracin, could be distinguished. On the other hand, inside the same chronological group it is very interesting to observe the different lead isotope ratios that appear for diverse types of glaze, transparent and tin-opacified, as in the samples from Albarracin;

two completely different sources of lead seem to have been used for producing these two types of lead glaze, and something similar could be inferred for Islamic samples from Zaragoza, although the low number of analyses does not allow one to draw definite conclusions in this case. This distinction could be related to the use of a type of commercial lead for producing transparent high-lead glazes and another, different, ceramic ingredient for tin-opacified glazes, probably purer or linked to the requirement of including tin in the glaze recipe. For instance, Abu'l Qasim's treatise explains the ingredients of one of the glaze preparations as being a mixture of white lead (lead carbonate?) and tin and the consequent joint production of tin and lead oxide for use in white glazes (Allan 1973). Also documented are different forms in which lead could be supplied to potters, and it is known that, in the 16th century AD, lead acquired for producing glazed cooking pottery was much cheaper than lead for tin-glazed ceramics (Alvaro 2002). In the 18th century AD the Ordinances of the Fraternities forbade Teruel potters from 'mixing common and fine lead glaze for making fine ceramic ware' (Alvaro 2002). Therefore, two different ceramic lead ingredients seem to be used for glazing pottery.

In order to extract as much information as possible, we can compare the glaze data with other published data from different lead sources in the Iberian Peninsula, situated in the south of the peninsula (the south-east: Cabo de Gata, Sierra de Almagrera, Sierra Gador, Sierra Alhamilla, Cartagena, Mazarrón; the central south: Alcudia Valley, Linares and Los Pedroches; and the south-west: Rio Tinto, Alnazcollar and others) (Arribas and Tosdal 1994, Marcoux 1998, Santos *et al.* 2004, Stos-Gale *et al.* 1995), in the north (Santander, Pais Vasco and Sierra la Demanda) (Velasco *et al.* 1996) and in the north-east (Cataluña) (Fig. 6.5) (Velasco *et al.* 1996, Canals and Cardellach 1997). Also some lead ores from the Catalonian coastal ranges in the Bellmunt area, mines exploited in Antiquity, were analysed by ICP-QMS (Table 6.4). All these important lead sources exploited in the Iberian Peninsula suggest the provision of lead to potters. Moreover, the western Mediterranean is recognized as the most likely lead source during some periods for producing glazes in areas as far away as Egypt (Wolf *et al.* 2003).

Sample	$^{206}Pb/^{204}Pb \pm sd$	$^{207}Pb/^{204}Pb \pm sd$	$^{208}Pb/^{204}Pb \pm sd$	$^{207}Pb/^{206}Pb \pm sd$	$^{208}Pb/^{206}Pb \pm sd$
Regia mine	18.319 ± 0.041	15.727 ± 0.056	38.818 ± 0.106	0.8585 ± 0.0011	2.1190 ± 0.0013
Min.1	18.430 ± 0.007	15.737 ± 0.014	38.650 ± 0.036	0.8539 ± 0.0007	2.0972 ± 0.0015
Linda Mariquita mine	18.357 ± 0.028	15.659 ± 0.023	38.459 ± 0.041	0.8520 ± 0.0001	2.0910 ± 0.0009
Min.2	18.364 ± 0.021	15.718 ± 0.032	38.631 ±0.096	0.8559 ± 0.0010	2.1036 ± 0.0029

Table 6.4
Lead isotope data for different lead ore samples (mean value and standard deviation, n=3)

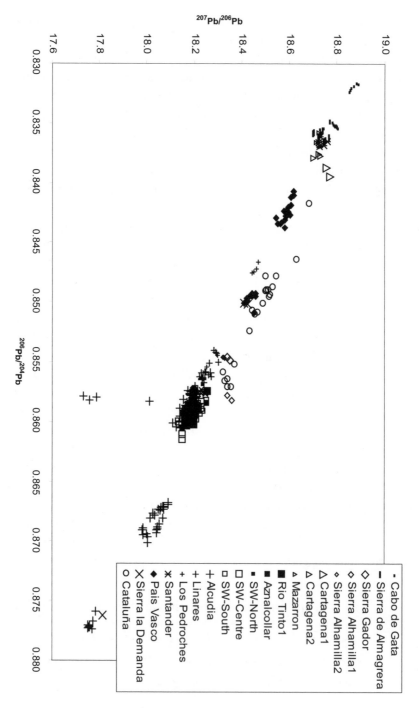

Fig. 6.5
Published lead isotope ratios for Iberian lead ores (^{206}Pb/^{204}Pb vs. ^{207}Pb/^{206}Pb)

If the glaze data are included in a more extended plot (Fig. 6.6), we can see that some glazes appear next to lead isotope ratios from the central south and the south-west, such as Rio Tinto. But the characteristic groups explained above (transparent high-lead glazes from Albarracin and tin glazes from Zaragoza) are near to the lead ore data of the central south (Linares or Los Pedroches), south-east (Sierra Gador or Sierra Alhamilla) or even north-east (Cataluña). The tin-glaze group from Albarracin does not appear to be definitely linked to a specific lead ore area. The same proximity is shown in the other bivariant plot. If we also include the lead ores analysed by ICP-QMS (Fig. 6.6), the use of lead coming from Catalonian mines might be strongly suggested for samples of Zaragoza glazes or transparent glazes from Albarracin.

It seems that, for the Islamic period, Zaragoza could have used lead extracted from the Catalonian coastal area for producing glazes. This would not be surprising, as this area belonged to the Zaragozan kingdom during that period. However, the Albarracin kingdom could have had other commercial relations with the kingdoms of the south. During the 14th to 16th centuries AD, in the Christian Aragon kingdom, Zaragoza could have maintained trade relations with the coast. Teruel has always been an area with more extensive commercial contacts with Valencia or Castille then with the southern lead sources.

Assignment of samples to an individual mine or area, however, is a very difficult and almost impossible task because of the likelihood of the lack of some Iberian ore sources and the possibility of overlapping lead isotope ratios if more new mines are included.

Conclusions

In spite of the low number of samples, some preliminary conclusions can be extracted. The performance of ICP-QMS is good enough to provide lead isotope-ratio determination with sufficient precision to differentiate between lead provenance from different localities and periods. The sample preparation is carried out by a simple methodology and does not imply tedious procedures of the separation of lead from the matrix. Moreover, the amount of the sample required is small. These first results for lead isotope ratios of glazes produced in medieval and post-medieval periods in the Iberian Peninsula show the possibility of establishing a manufacturing pattern, for different areas and periods, although a higher number of analyses is required. Contrary to what was shown in studies of glazes in other production centres (Wolf *et al.* 2003), in this study a distinction between lead sources chosen for producing different glaze types (transparent or tin-opacified)

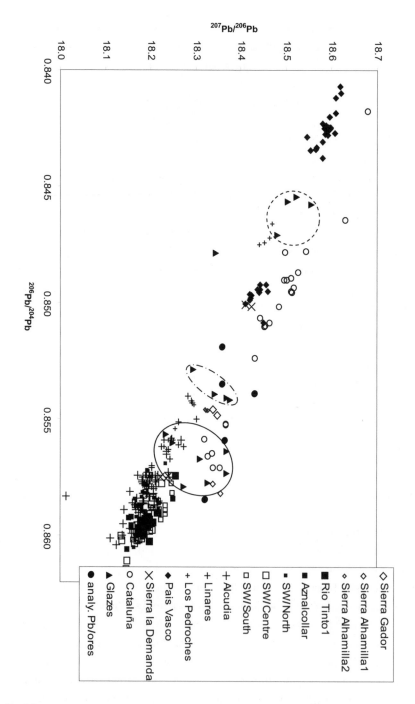

Fig. 6.6
Lead isotope ratios for Iberian lead ores and glazed pottery samples (^{206}Pb/^{204}Pb vs. ^{207}Pb/^{206}Pb)

can be observed in the glazes from Albarracin, and probably also from Zaragoza. The two types of glazes produced in these sites can be distinguished by lead isotope analysis.

Acknowledgements

This study was financially supported by CTPP03/2005 research project of the 'Comunidad de Trabajo de los Pirineos' (Aragon–Catalunia-Aquitaine) and Diputación General de Aragón (DGA). Samples were provided by the Museum of Zaragoza, the Museum of Teruel and the City Council of Zaragoza. We want also to thank Maurice Picon and Julian Henderson for providing samples from Pechina and Raqqa, and Ignacio Montero for supplying lead-ore samples and TIMS data.

References

J. W. Allan, 1973, Abu'l-Qasim's treatise on ceramics, Iran, IX, 111-120.

I. Alvaro Zamora, 2002, Cerámica aragonesa, 1, Ibercaja.

A. Arribas, Jr., R. M. Tosdal, 1994, Isotopic composition of Pb in ore deposits of the Betic Cordillera, Spain: Origin and relationship to other European deposits, Enomonic Geology, 89, 1074-1093.

M. Beltrán Lloris, 1990, Guía de la cerámica romana, Zaragoza.

A. Canals, E. Cardellach, 1997, Ore lead and sulphur isotope pattern from the low-temperature veins of the Catalonian Coastal Ranges (NE Spain), Mineralium Deposita, 32, 243-249.

J. Casas i Genover, J. Merino i Serra, 1990, Troballes de céramica vidriada d'època romana a les comarques costaneres de Girona, CYPSELA, VIII, 139–155.

E. Cooper, 1998, A History of world pottery, B.T. Bastford.

A. López Mullor, 1978, Cerámica vidriada romana, Información Arqueológica, 27–28, 68–74.

N. H. Gale, Z. Stos-Gale, 2000, Lead isotope analyses applied to provenance studies, in: E. Ciliberto, G. Spoto (eds.) Modern Analytical Methods and Art and Archaeology, 155, John Wiley & Sons, 503-584.

J. A. Habicht-Mauche, S. T. Glenn, H. Milford, A. Russell Flegal, 2000, Isotopic tracing of prehistoric Rio Grande glaze-paint production and trade, Journal of Archaeological Science, 27, 709-713.

J. Herderson, J. A. Evans, H. J. Sloane, M. J. Leng, C. Doherty, 2005, The use of oxygen, strontium and lead isotopes to provenance ancient glasses in the middle East, Journal of Archaeological Science, 32, 665-673.

E. Marcoux, 1998, Lead isotope systematics of the giant massive sulphide deposits in the Iberian Pyrite Belt, Mineralium Deposita, 33, 45-58.

J. Molera, M. García-Vallés, T. Pradell, M. Vendrell, 1996, Hispano-Moresque productions of the fourteenth-century workshop of the Testar del Molí (Paterna, Spain), Archaeometry, 38, 67-80.

J. Molera, M. Vendrell-Saz, J. Pérez-Arantegui, 2001, Chemical and textural characterization of tin glazes in Islamic ceramics from eastern Spain, Journal of Archaeological Science, 28, 331-340.

J. Pérez-Arantegui, M. I. Uruñuela, M. P. Lapuente, J. R. Castillo, 1995, Study of Roman lead glazing technology between 1st and 2nd centuries AD by Scanning Electron Microscopy, in: 4th European Ceramic Society Conference, Cultural Ceramic Heritage/3rd European Meeting on Ancient Ceramics, Riccione, Italy.

J. Pérez-Arantegui, M. P. Lapuente, 2003, Las técnicas de producción de cerámicas en los talleres islámicos de Zaragoza (España), in: Ch. Bakirtzis (ed.) La céramique médiévale en Méditerranée, Édition de la Caisse des Recettes Archéologiques, 375-380.

J. Pérez-Arantegui, J. Ortega, C. Escriche, in press, The Hispano-Moresque tin-glazed ceramics produced in Terruel, Spain: A technology between two historical periods, 13th-16th c. AD. In From Mine to Microscope - Studies in honour of Mike Tite. University College London Press.

C. Pomiès, A. Coherie, C. Guerrot, E. Marcoux, J. Lancelot, 1998, Assessment of the precision and accuracy of lead isotope ratios measured by TIMS for geochemical applications: example of massive sulphide deposits (Rio Tinto, Spain), Chemical Geology, 144, 137-49.

J. F. Santos Zalduegui, S. García de Madinabeitia, J. I. Gil Ibarguchi, F. Palero, 2004, A lead isotope database: the Los Pedroches –Alcudia area (Spain). Implications for archaeometallurgical connections across Southwestern and Southeastern Iberia, Archaeometry, 46, 625-634.

R. W. Sheets, 1999, Release of heavy metals from European and Asian porcelain dinnerware, The Science of Total Environment, 212, 107-113.

Z. Stos-Gale, N. H. Gale, J. Houghton, R. Speakman, 1995, Lead isotope data from the Isotrace Laboratory, Oxford: Archaeometry database 1, ores from the Western Mediterranean, Archaeometry, 37, 407-415.

M. S. Tite, I. Freestone, R. Mason, J. Molera, M. Vendrell-Saz, N. Wood, 1998, Lead glazes in antiquity- methods of productions and reasons for use, Archaeometry, 40, 241-260.

F. Velasco, A. Pesquera, J. M. Herrero, 1996, Lead isotope study of Zn-Pb deposits associated with the Basque-Cantabrian basin and Paleozoic basement, Northern Spain, Mineralium Deposita, 31, 84-92.

S. Wolf, S. Stos, R. Mason, M. S. Tite, 2003, Lead isotope analyses of Islamic pottery glazes from Fustat, Egypt, Archaeometry, 45, 405-420.

K.A. Yener, E. V. Sayre, H. Ozbal, E. C. Joel , I.L. Barnes, R.H. Brill, 1991, Stable lead isotope studies of Central Taurus ore sources and relates artefacts from Eastern Mediterranean Chalcolithic and Bronze Age Sites, Journal of Archaeological Science, 18, 541-577.

PLS Regression to Determine Lead Isotope Ratios of Roman Lead Glazed Ceramics by Laser Ablation TOF-ICP-MS

Marc S. Walton

Introduction

Inductively coupled plasma mass spectrometry (ICP-MS) has increasingly become the favoured technique in recent years for the measurement of lead isotope ratios (eg., Baker *et al.* 2006, Reuer *et al.* 2003, Woodhead 2002, Clayton *et al.* 2002, Ehrlich *et al.* 2001, Wannemacker *et al.* 2000, Horn *et al.* 2000, Rehkämper and Halliday 1988, Heumann *et al.* 1998). In comparison to thermal ionization mass spectrometry (TIMS), the benchmark instrument used for Pb isotope measurements, ICP-MS does not require laborious sample preparation but can achieve adequate precision (>0.1% RSD) with careful experimental design. The utility of ICP-MS has become particularly evident in archaeological applications where instruments are often coupled to lasers allowing for ablative sample removal on a spatially resolved micron scale. The major advantage of using a laser is that the isotopic ratio information may be obtained very rapidly with little to no sample preparation (Baker *et al.* 2006, Ponting *et al.* 2003). The speed of laser ablation ICP-MS means, in terms of throughput, that many more samples may be analysed by this technique in comparison to by TIMS. As a consequence, larger datasets of isotopic information can be compiled by ICP-MS, thus making isotopic analysis a less specialized method in determining the provenance of archaeological materials.

Of course, there are some inherent limitations in using laser ablation ICP-MS to obtain accurate and precise isotopic measurements. Elemental fractionation can occur during laser ablation yielding inaccurate results (Liu *et al.* 2004). Also, the loading of the plasma with complex sample matrices (i.e., those that have not undergone special chemical separation techniques to pre-concentrate Pb) can induce the unfavourable space-charge effects that cause mass bias (Rehkämper and Mezger 2000). Finally, with low-resolution instrumentation, tailing from adjacent masses may cause an over-estimation of the true intensity of the isotope of interest (Reuer *et al.* 2003). Therefore, to obtain accurate and precise isotope ratios with plasma-based instrumentation coupled to laser ablation sample introduction, the effects of fractionation, mass bias, and peak tailing require evaluation and correction.

With solution nebulization techniques of sample introduction, mass bias is often corrected by internal normalization of the radiogenic isotope of interest using a stable isotope ratio of fixed value (i.e., $^{205}Tl/^{203}Tl$ for Pb isotopes). By running standards containing both stable and radiogenic ratios of known value, the relationship between the mass bias of the stable and radiogenic ratio can be characterized and then applied to any unknown samples. When analysing solids with laser ablation, however, not all unknown samples have measurable $^{205}Tl/^{203}Tl$ available to correct for the values of the Pb isotope ratios. Furthermore, the effects of fraction caused by laser ablation may alter the relationship between the stable and radiogenic isotope ratios. Therefore, other methods must be used to ensure accurate isotope ratio measurements such as bracketing unknown samples with standards of known isotopic composition (Baker et al. 2006). This method suffers from the problem of loading a plasma with the dry aerosols produced by laser ablation. In these situations, where complex sample matrices enter the plasma, the ionization process can cause differential mass biases (depending on sample composition) and polyatomic interferences to form (Rehkämper and Mezger 2000).

In this paper, a relatively new variety of ICP-MS with a time-of-flight (TOF) mass analyser is discussed (Yang et al. 2005, Willie et al. 2005, Kozlov et al. 2003, Leach and Hieftje 2001). The TOF instrument is quite different from TIMS based instrumentation, which measures mass by magnetic deflection in a flight tube. TOF-ICP-MS (hereafter TOF) instead employs a single discrete dynode detector to measure the mass of an ion, which is determined by the time required for the ion to traverse a flight path of 1.4 metres (heavy masses take a longer time than the light masses to travel from one end of the flight tube to the other). Although TOF is not as well suited to isotopic measurements as multi-collector and magnetic sector mass spectrometers (both plasma-based instruments) as the peaks are not flat topped and there is inferior abundance sensitivity, it has been shown, nonetheless, that TOF instruments are capable of reproducing an isotope ratio precision of 0.04% relative standard deviation (RSD) (Willie et al. 2005). This value of precision is more than sufficient for archaeological tracer studies. Furthermore, TOF instruments have significant advantages over these other standard ICP-MS instruments (e.g. quadrapole mass analysers and sector field instruments that sequentially scan the mass range) due to the fact that an entire mass spectrum is obtained for every ion-gating event. Because this essentially synchronous signal counts over the entire mass spectrum, there is a high correlation of noise for all isotopes measured and no spectral skew, thus allowing for the accurate measurement of almost the entire periodic table of isotopes in a single experiment. Therefore, this technique is ideally suited to experimental situations where a large number of isotopes (often more than fifteen) are monitored in fast transient signals produced by solid sample introduction methods like laser ablation.

The aim of this study was to develop an analytical approach to measuring Pb isotope ratios with the TOF instrument using laser ablation sample introduction. As a test case for the routine analysis of isotope ratios with this instrument, a series of twenty-one glazes from early Roman ceramics were analysed. Since seven of these glazes had been run previously by TIMS, this allowed for a calibration routine to be developed based on their Pb isotopic composition. Also from the results of this calibration, the precision and accuracy achieved by the TOF instrument were compared to the TIMS generated values.

Experimental

PLS MODELING

To take full advantage of the simultaneous and multiple element capabilities of the TOF instrument, a partial least squares regression (PLS) calibration model was generated to develop an empirical representation of Pb isotope mass bias and peak tailing. PLS is related to principal component analysis (PCA) in that the technique linearly transforms a large number of variables into components that reflect the greatest variance in the dataset (Zhu *et al.* 1997). However, instead of using the variance to separate the dataset into subsets (such as in PCA), in PLS the variance is used to describe a set of predicted variables in terms of a matrix of observed variables to produce a linear calibration model. In ICP-MS applications, this multivariate approach has been applied previously in evaluating spectral interferences caused by sodium, calcium, chlorine and sulphur (Grotti *et al.* 1999) and in developing a calibration model for the determination of rare earth elements (Zhu *et al.* 1997). Since the fundamentals of PLS have been presented elsewhere, only brief details outlining our procedure will be presented here.

For this study, the Y (nxm) matrix of values holds m lead isotope ratio signatures of n multi-element lead isotope standards, and the X (nxp) matrix holds the p observed counts of n isotopes obtained by laser ablation TOF-ICP-MS. The X matrix of values includes the Pb isotope ratios themselves, isotopes adjacent to the lead isotopes (that could cause tailing ineterferences), major element concentrations (to account for space charge effects), and known isobaric intereferences to Pb isotopes (see Table 7.1 for values included in the model). In broad terms, the decomposition of the X and Y matrices is as follows:

Isotope Ratios	$^{208}Pb/^{206}Pb$, $^{207}Pb/^{206}Pb$, $^{206}Pb/^{204}Pb$
Isobaric Interferences	^{202}Hg for $^{204}Pb+^{204}Hg$ and $^{164}Dy^{40}Ar$, $^{168}Er^{40}Ar$
Matrix Elements	Na, Mg, Al, Si, Ca, K, Fe, Ti
Adjacent Isotopes (peak tailing)	^{203}Tl, ^{205}Tl, ^{209}Bi

Table 7.1
Variables input into PLRS model

$$X = TP \tag{1}$$

$$Y = UQ \tag{2}$$

where T and U are the score matrices, and P and Q are the loading matrices.

As a linear equation, the matrix can be written as follows for a given isotope ratio

$$Y_{ratio} = b_0 + \sum_{i=1}^{n} b_i x_i + \sum_{i=1}^{n} b_i x_i^2 + \sum_{i \neq j; j=1}^{n} b_i x_i x_j \tag{3}$$

where Y_{ratio} is the corrected isotope ratio, x_i is the isotope ratio or normalized isotope counts which are the matrix values, and b_i is the coefficient determined by a least squares fit between the scores of the X and Y matrix blocks. If, hypothetically, six variables adjacent to $^{208}Pb/^{206}Pb$ isotope ratio were used to determine the proper value of this ratio, then the equation would appear as follows:

$$Y_{208Pb/206Pb} = b_0 + b_1(^{208}Pb/^{206}Pb) + b_2(^{207}Pb/^{206}Pb) + b_3(^{206}Pb/^{204}Pb) + b_4(^{209}Bi) + b_5(^{205}Tl) + b_6(^{203}Tl) \tag{4}$$

where the values in brackets are either measured isotope ratios or isotopic counts normalized to an internal standard.

STANDARDS

Calibration was performed by analysing seven Roman lead-glazed ceramics, for which lead isotope ratios were determined by TIMS (Walton 2004), and a lead silicate glass formed from NIST SRM 981.

The lead silicate glass (hereafter 981 glass) was prepared by disolving NIST SRM 981 lead wire (National Institue of Standards and Technology, USA) in concentrated high purity HNO_3 (Optima, Fischer Scientic, USA). The solution of NIST SRM 981 was allowed to evaporate to dryness forming solid lead nitrate

Pb(NO$_3$)$_2$ crystals. The Pb(NO$_3$)$_2$ was intimately combined with a high silicate ceramic powder (JRRM 121, Certified Reference Materials for X-ray Fluoresence Analysis of Refractories, The Technical Association of Refractories, Japan) to produce a mixture with elemental lead at the 61% atomic weight level which was chosen to be similar in composition to the Roman lead glazes. The lead and ceramic standard mixture was packed into an alumina crucible and ramped at a rate of 10°C/min to 850°C in a muffle furnace for a dwell time of two hours (Barnstead 47900, Barnstead International, USA). The resulting lead silicate glass was then removed from the furnace and allowed to cool to room temperature. There was no further preparation of the 981 glass prior to analysis.

SAMPLES

All of the lead-glazed ceramics had previously been embedded in epoxy resin, polished in cross-section, and carbon coated in preparation for analysis by electron microprobe (Walton 2004). To prepare the sections for analysis by ICP-MS, the carbon coating of the samples was removed by polishing which also revealed a fresh surface.

INSTRUMENTATION

Measurements were made using a GBC Optimas 9500 TOF-ICP-MS (GBC Scientific Equipment Pty. Ltd, Australia) coupled to a New Wave UP 213 laser ablation system (New Wave Instruments, Freemont, CA). Optimization of the ion beam and peak shape was performed daily by monitoring [208]Pb while ablating the NIST610 silicate glass standard (National Institue of Standards and Technology, USA). Typical instrument settings can be seen in Table 7.2.

Data acquisition was performed with an AP100 Acquiris digital signal averager card (Acquiris, Geneva, Switzerland) which operates at 0.5 gigasamples per second providing a sampling interval of 2ns with the TOF instrument (Willie *et al.* 2005). On board data signal processing was performed with the Noise Suppressed Accumulation (NSA) function that sets a voltage threshold above the random noise which is superimposed on the baseline voltage. In normal operation, before an isotope signal is summed on the Acquiris card, the isotope peak must exceed the set threshold. Once an isotope peak has exceeded the threshold value, counts from the baseline voltage are added back into the peak for total baseline to peak integration. For acquisition of isotope ratios, this threshold was set to a low value (one analogue to digital conversion, ADC) which included a portion of random noise in the spectrum. This was done to avoid errors associated with incorrect baseline to peak summation which had been observed previously when setting the NSA to its full value (~ 3 ADC).

Laser Ablation:	
Wavelegth	213nm
Spot Size	55mm
Fluence	10 J cm^{-2}
Rep Rate	5 Hz
He gas flow rate	0.3 l min^{-1}

ICP Source:	
RF power	750 W
Sample Ar gas flow	0.9 l min^{-1}
Auxiliary Ar gas flow	0.5 l min^{-1}
Plasma Ar gas flow	10.0 l min^{-1}

Table 7.2
Typical instrument operating conditions

In order to reduce the gross effects of fractionation caused by laser ablation, the large non-stoichiometric particles were removed from the ablation aerosol by online filtering using a glass wool plug (Guillong and Günther 2002).

DATA REDUCTION PROCEDURES

Transient signals were detected in fifty-second windows triggered by a laser ablation event of twenty seconds' duration. Therefore, for every transient produced there were twenty seconds of signal and thirty seconds of background collected (Fig. 7.1). Over the fifty-second window, 128 complete spectra defined the transient signal producing an integration time of 0.26s/spectrum. The transient signal data was compiled in Microsoft Access and then transferred to Microsoft Excel for data reduction.

Prior to modelling the data by PLS, on-peak background windows of equal size to the signal window were subtracted to produce a background corrected signal. Also, as shown in Fig. 7.2, the uncalibrated Pb isotope ratios were calculated by linearly regressing the individual isotope data points using the LINEST function in Microsoft Excel. The standard error of the slope of this regression provided an estimation of the internal standard error of the lead isotope ratio measurement. After linear regression analysis, the PLS model was then implemented using XLSTAT add-in for Microsoft Excel (Addinsoft, France).

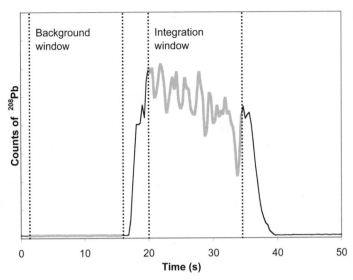

Fig. 7.1
Transient signal produced by laser ablation. Background windows were measured before each transient peak. The integration window was set at the same length as the background. Each transient peak was composed of 128 mass spectra over a fifty-second period

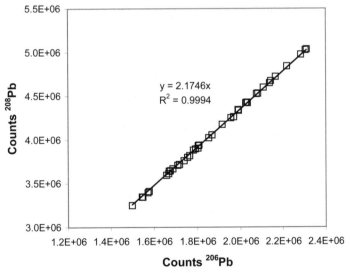

Fig. 7.2
Linear regression performed on the measurements of the ^{208}Pb and ^{206}Pb isotopes from a single transient peak of NIST 981 after background subtraction. The slope provides the isotope ^{208}Pb/^{206}Pb ratio before mass bias and drift corrections. Prediction of error of the slope is calculated from the regression. In the above case, the slope standard error was 0.001 or 0.05% RSD. This calculation of error provides an estimate of the internal error of the isotope measurement

Results

CALIBRATION

Linear regression of the transient isotope signals (shown in Fig. 7.2) produced uncorrected isotope ratios as determined from the slope of the line with an estimated internal error of approximately 0.05% RSD. This precision is close to the limiting precision obtained previously using a similar make of instrument (Willie *et al.* 2005). After obtaining at least six replicates of uncorrected isotope ratios for each standard, the PLS model was produced. Fig. 7.3 shows the result of PLS for the $^{207}Pb/^{206}Pb$ ratio where the values generated by TIMS analysis are compared to the values predicted by the model. In Fig. 7.3, the dotted line is the mean of the model while the solid lines show the 95% confidence intervals of the predicted fit that provides an estimation of the global accuracy of the model. Based on this statistic, the ratio predicted by PLS was found to fall within an interval of $\pm$ 0.003 of the TIMS-derived isotope ratio measurements. Another useful statistic measured was the correlation coefficient (r^2) that establishes the linearity of the model in the range of values measured. The r^2 value of the $^{207}Pb/^{206}Pb$ ratio (shown in Fig. 7.3) was found to be 0.998. This value is similar to the correlation coefficients obtained for the $^{208}Pb/^{206}Pb$ and $^{206}Pb/^{204}Pb$ ratios, which were 0.913 and 0.985 respectively. These correlation values demonstrate the goodness of fit between the TIMS-generated values and the predicted values, thus signifying that the variables were correctly modelled by PLS.

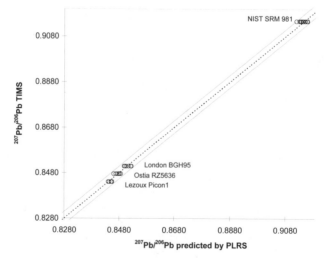

Fig. 7.3
Results of the PLRS model comparing TIMS-based measurements (Y-axis) to those predicted by the model

As a secondary check of the calibration model, a series of seven lead-glazed sherds with known isotope ratios were analysed (inclusive of the three glazed sherds used in the calibration). Fig. 7.4 shows the run sequence of these glazes (monitoring the $^{208}Pb/^{206}Pb$ ratio) where each glaze was analysed in six replicates. Each sample analysis was also bracketed between analyses of the standard 981 glass. The running averages of the 981 glass for the $^{208}Pb/^{206}Pb$, $^{207}Pb/^{206}Pb$, and $^{206}Pb/^{204}Pb$ ratios were respectively 2.168 ± 0.001, 0.915 ± 0.001, and 16.93 ± 0.02 (where the certified values are 2.1681, 0.91464, and 16.9371 for $^{208}Pb/^{206}Pb$, $^{207}Pb/^{206}Pb$, and $^{206}Pb/^{204}Pb$). This represents an overall replicate precision of approximately 0.1% RSD for each of the isotope ratios in this run sequence. This precision on the 981 glass is comparable to routine analysis of NIST 981 on a TIMS instrument previously dedicated to archaeological materials (Stos-Gale *et al*. 1995).

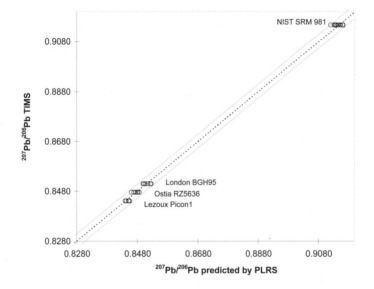

Fig. 7.4
Plot shows a single session of data acquisition where each sample analysis is bracketed by a measurement of a glass made from NIST SRM 981. The running average of NIST 981 (solid line through 981 data points) was calculated after drift correction. External precision (dotted lines) for NIST 981 through the course of this run was found to be ± 0.001, which represents a relative standard deviation of 0.07% at 2 sigma. This precision is better than that typically obtained by TIMS for archaeological tracer studies. As demarcated by a solid line, every six data points represents a new lead glaze for a total of seven sherds in this run

In Table 7.3 the average values for each of the glazed sherds analysed by TOF are
compared to the isotope ratios obtained by TIMS. From these analyses it can be
seen that good accuracy is achieved for each of the ratios and the reproducibility
is slightly inferior to lead isotope ratios measured by laser ablation on a multi-
collector instrument (Baker *et al.* 2006). Fig. 7.5 shows this comparison graphically.
It can be seen from these comparison plots that the measured TIMS standard error
(which is less than the width of each data point) is actually much smaller than that
obtained by TOF. The error derived from TIMS is approximately 0.01% RSD for
each of the isotopes measured whereas the TOF yields precision of between 0.05
and 0.1% RSD on repeat measurements. Despite this difference in precision, the
TOF data are still good enough to be used for studies on archaeological provenance.

Sample		$^{208}Pb/^{206}Pb$	$^{207}Pb/^{206}Pb$	$^{206}Pb/^{204}Pb$
Diana 5571a	TIMS	2.0779 ± 0.0001	0.83852 ± 0.00002	18.696 ± 0.001
	TOF MEAN	2.0777	0.8391	18.69
	RSD %	0.005	0.009	0.04
London LCT84b	TIMS	2.0859 ± 0.0001	0.84558 ± 0.00001	18.519 ± 0.001
	TOF MEAN	2.0839	0.845	18.54
	RSD %	0.02	0.16	0.05
London LCT84a	TIMS	2.08669 ± 0.00006	0.84962 ± 0.00002	18.519 ± 0.001
	TOF MEAN	2.0865	0.8488	18.50
	RSD %	0.03	0.09	0.1
Ostia RZ5636	TIMS	2.0911 ± 0.0001	0.84753 ± 0.00002	18.467 ± 0.001
	TOF MEAN	2.0915	0.8476	18.53
	RSD %	0.04	0.09	0.14
Lezoux Picon 1	TIMS	2.08485 ± 0.00006	0.84408 ± 0.00001	18.535 ± 0.001
	TOF MEAN	2.086	0.8446	18.53
	RSD %	0.05	0.08	0.14
Diana C135-96	TIMS	2.0836 ± 0.0001	0.84126 ± 0.00001	18.634 ± 0.001
	TOF MEAN	2.082	0.8406	18.63
	RSD %	0.07	0.16	0.13
London BGH95	TIMS	2.09777 ± 0.0001	0.85096 ± 0.00002	18.407 ± 0.001
	TOF MEAN	2.095	0.8510	18.40
	RSD %	0.07	0.08	0.08

Table 7.3
Comparison of TOF Pb isotope ratios with those obtained by TIMS

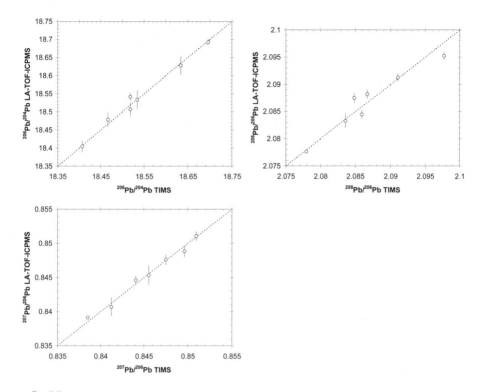

Fig. 7.5
Bivariate plots comparing TIMS values to those obtained by LA-TOF-ICPMS for the $^{208}Pb/^{206}Pb$, $^{207}Pb/^{206}Pb$, $^{206}Pb/^{204}Pb$ ratios. The dotted line is the 1:1 ratio

LEAD-GLAZED SAMPLES

After establishing the calibration model, fourteen Roman glazed sherds from the 1st century AD were analysed to access similarities and differences in the origin of the lead used in their manufacture. All of the sherds were previously determined to have been made in Italian workshops on the basis of their similar mineral assemblages and calcium rich compositions (Walton 2004). However, as shown in Table 7.4, the sherds were excavated from a variety of different locations.

Table 7.4 is divided into two sections: one containing yellow glazes as denoted by a 'y' superscript, and the other containing green glazes as denoted by a 'g' superscript. The green colour is from a copper-based colorant that was added to the glaze. The yellow is due to iron that is likely to have migrated into the glaze by reaction with the ceramic body. Also shown in Table 7.4 are the lead isotope ratios of the glazes, together with the RSDs associated with each analysis. It may be seen that, with few exceptions, the precision achieved in these analyses surpasses the 1% RSD limit that has been deemed acceptable for archaeological tracer studies. As shown in Fig. 7.6, both the yellow and green glazes have isotope ratios that match Spanish ore sources which have been recently characterized (Santos Zalduegui *et al.* 2004).

Sample	Findspot	$^{208}Pb/^{206}Pb$	%RSD	$^{207}Pb/^{206}Pb$	%RSD	$^{206}Pb/^{204}Pb$	%RSD
V69^y	Celsa, Spain	2.0987	0.287	0.8467	0.485	18.326	0.595
V27^y	Celsa, Spain	2.0749	0.002	0.8353	0.157	18.722	0.054
V59^y	Celsa, Spain	2.0896	0.014	0.8478	0.001	18.436	0.022
V83^y	Celsa, Spain	2.0936	0.015	0.8545	0.140	18.335	0.070
SWA81^y	London, England	2.0940	0.143	0.8447	0.250	18.403	0.297
SH74^y	London, England	2.1140	0.015	0.8722	0.278	17.935	0.113
RZ5670^y	Ostia, Italy	2.0886	0.008	0.8479	0.002	18.450	0.012
Picon4^y	Lezoux, France	2.1039	0.161	0.8544	0.087	18.201	0.271
Camp1^y	Campania, Italy	2.1086	0.425	0.8566	0.256	18.123	0.724
Camp3^y	Campania, Italy	2.0953	0.034	0.8525	0.045	18.327	0.066
Camp4^y	Campania, Italy	2.0938	0.006	0.8575	0.013	18.310	0.014
V69^g	Celsa, Spain	2.0886	0.018	0.8479	0.040	18.489	0.033
V27^g	Celsa, Spain	2.0807	0.009	0.8413	0.019	18.638	0.024
V59^g	Celsa, Spain	2.0863	0.045	0.8460	0.093	18.526	0.101
V83^g	Celsa, Spain	2.0976	0.018	0.8554	0.033	18.303	0.055
SWA81^g	London, England	2.0959	0.019	0.8541	0.038	18.347	0.047
SH74^g	London, England	2.0916	0.019	0.8461	0.055	18.422	0.047
RZ5670^g	Ostia, Italy	2.0876	0.055	0.8485	0.056	18.457	0.099
Picon4^g	Lezoux, France	2.0988	0.016	0.8562	0.031	18.262	0.046
Camp1^g	Campania, Italy	2.0959	0.010	0.8551	0.011	18.300	0.020
Camp2^g	Campania, Italy	2.0914	0.020	0.8507	0.084	18.390	0.058
Camp3^g	Campania, Italy	2.0944	0.001	0.8571	0.013	18.305	0.003
rz5574^g	Ostia, Italy	2.0951	0.033	0.8533	0.068	18.353	0.067
rz5637^g	Ostia, Italy	2.0869	0.089	0.8463	0.183	18.495	0.202
DGH86^g	London, England	2.0887	0.009	0.8480	0.018	18.487	0.022

Table 7.4
Roman lead-glazed sherds analysed by TOF

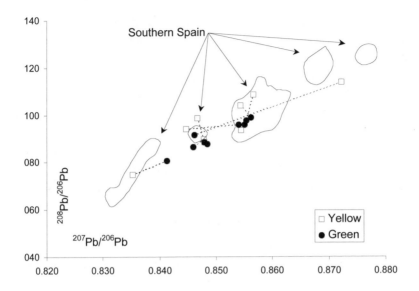

Fig. 7.6
Bivariate plot showing $^{207}Pb/^{206}Pb$ vs. $^{208}Pb/^{206}Pb$ ratios of Roman lead glazed sherds

Discussion

CALIBRATION

To be able to correct for mass bias it must be measured in an empirical manner since its occurrence in plasmas is not well understood on a fundamental level (Andrén 2004). Since PLS is a multi-variate modelling technique, it takes into account all the different behaviours of elements when they are ionized from slightly dissimilar sample matrices. Therefore it provides a robust route towards measuring mass bias in a plasma when internal standards are not available for normalization of the isotope ratios. It is also believed that PLS may provide the basis for a more fundamental understanding of mass bias. With PLS, the influences of different compositional variables can be assessed and, indeed, quantified (Grotti *et al.* 1999). Such investigations on mass bias with PLS will be part of our ongoing investigations of the TOF instrumentation.

When using PLS calibration models with plasma instrumentation, however, there are several caveats that should be applied to achieve accurate results.

i)   The standards used in the calibration must be properly matrix-matched to the samples. In the case of the lead glazes studied in this paper, actual archaeological glazes which had been previously analysed by TIMS were used to ensure matrix-matched conditions. Furthermore, the fabricated NIST 981 glass in this study was designed to have a similar matrix composition to the lead glazes. Having similar compositions between standards and samples effectively reduces the compositionally induced mass biases, as well as peak tailing, to constants. In fact, the PLRS model specifically measures these constants for each variable influencing the isotope ratio.

ii) The range of isotope ratios of the standards defines the linear range of the model. As a consequence, any sample isotope ratio measurement that falls outside the range of standard values is not likely to be properly characterized by the model, and therefore must be discarded.

iii) Because of temporal changes in plasma temperature, it is necessary to use a bracketing standard (such as the 981 glass used in this study) to protect against instrument drift.

If the above methods of minimizing the instrumental bias are taken into account, Pb isotope ratios of high accuracy and precision can be achieved, as is seen by the data presented in Tables 7.3 and 7.4.

MEASUREMENT OF LEAD ISOTOPE RATIOS OF ROMAN LEAD GLAZES

It can be seen in Fig. 7.6 that the isotope ratios of yellow and green glazes coming from the same sherd (denoted by dotted tie-lines) do not always possess similar isotopic compositions. The most notable isotopic difference between the yellow and green glazes on a single sherd may be seen with sample SH74. This sherd has a green glaze with a $^{207}Pb/^{206}Pb$ ratio of 0.8461 and a yellow glaze with the same ratio equal to 0.8722. This discrepancy in isotopic composition suggests that the lead used in the fabrication of these two colours comes from different sources. It is possible that the Cu colorant used in the green glazes could have contributed to the isotopic composition of the glaze if scrap leaded bronze was used. However, since Cu amounts to, at most, 3% by weight in these glazes (Walton 2004), mixing of a leaded bronze with the lead of the glaze should account for only a small change in the Pb isotopic signature of the glaze. Therefore, the contribution of lead from the copper colorant can be discounted in accessing the lead isotope ratios of these glazes.

It is not altogether surprising that lead from multiple sources may be found on a single vessel. It may be envisaged that in glazing these vessels a workshop would have two glaze suspensions available: one yellow and the other green. If one of these suspensions ran out, it could have been replaced by a suspension made from lead coming from a different isotopic source.

Of greater interest is the match of lead in the study with Spanish ore sources. Previously, the isotopic ranges demonstrated by the sherds in this study would have been attributed to British sources such as the Mendips and Cumbria (Butcher and Ponting 2005, Stos-Gale *et al*. 1995). However because of the extensive silver mining activities in Southern Spain during the 1st century AD, especially at Rio Tinto (Craddock1995), the ore sources identified by Santos Zalduegui *et al*. (2004) seem to be a logical source of the recycled litharge that would have been used in glazing ceramics.

As part of our continuing investigations of Roman lead-glazed ceramics, the lead isotope ratios measured in this study will be compared to the trace element values in these glazes. It is hoped that these trace elements will help to confirm our understanding of ore sources and use/re-use of lead in the Roman world.

Conclusion

Laser ablation TOF-ICP-MS has been used to perform lead isotope ratio analysis of Roman glazes. When comparing the values obtained by this technique with previous TIMS analyses, it may be seen that the TOF-ICP-MS can generate the accuracy and precision necessary for archaeological applications. The accuracy and precision were achieved by modelling a variety of interference and mass bias effects with multiple linear regression. There is, however, considerable potential for the improvement of this modelling technique. Most importantly, it is desirable to limit the number of parameters used in the model to those causing only significant changes to the isotope measurement. By doing this, the cause of mass bias and/ or peak tailing may be isolated and specifically corrected. This will be the major focus of ongoing investigations to obtain isotope ratios with the TOF instrument.

Although it is clear that more work is needed to clarify the origin of the Roman glazed sherds examined in this study, it is plausible that the lead used in their manufacture comes from ore sources in Southern Spain. Trace element analysis should be able to confirm whether Spanish ore sources are possible for these glazes, and this will be the subject of forthcoming work.

Acknowledgements

The author gratefully acknowledges Karen Trentelman, of the Getty Conservation Institute, and Stefan Buerger, of Oak Ridge National Laboratories (USA), for reading and commenting on an early version of this manuscript. Bill Balsanek, of GBC Scientific Instruments, also greatly aided in the technical understanding of the TOF-ICP-MS instrument.

References

H. Andrén, 2004, Studies of Artificial Mass Bias in Isotopic Measurements by Inductively Coupled Plasma Mass Spectrometry, Unpublished Ph.D. Thesis, Lulea University of Technology.

J. Baker, S. Stos, T. Waight, 2006, Lead Isotope Analysis of Archaeological Metals by Multiple-Collector Inductively Coupled Plasma Mass Spectrometry, Archaeometry, 48, 45-56.

K. Butcher, M. Ponting, 2005, The Roman denarius under the Julio-Claudian emperors: mints, metallurgy, and technology, Oxford Journal of Archaeology, 24, 163-197.

R. Clayton, P. Andersson, N.H. Gale, C. Gillis, M.J. Whitehouse, 2002, Precise determination of the isotopic composition of Sn using MC-ICP-MS, Journal of Analytical Atomic Spectrometry, 17, 1248-1256.

P. Craddock, 1995, Early Metal Mining and Production, Edinburgh University Press.

S. Ehrlich, Z. Karpas, L. Ben-Dor, L. Halicz, 2001, High precision lead isotope ratio measurements by multicollector-ICP-MS in variable matrices, Journal of Analytical Atomic Spectrometry, 16, 975-977.

M. Grotti, C. Gnecco, F. Bonfiglio, 1999, Multivariate quantification of Spectroscopic interferences casued by sodium, calcium, chlorine, and sulfur in inductively coupled plasma mass spectrometry, Journal of Analytical Atomic Spectrometry, 14 , 1171-1175.

M. Guillong, D. Günther, 2002, Effect of particle size distribution of ICP-induced elemental fractionation in laser ablation - inductively coupled plasma - mass spectrometry, Journal of Analytical Atomic Spectrometry, 17, 831-837.

I. Horn, R. Rudnick, W.F. McDonough, 2000, Precise elemental and isotope ration determination by simulatneous solution nebulization and laser ablation-ICP-MS: application to U-Pb geochronology, Chemical Geology, 167, 405-425.

K. Heumman, S. Gallus, G. Rädlinger, J. Vogl, 1998, Precision and accuracy in isotope ratio measurements by plasma source mass spectrometry, Journal of Analytical Atomic Spectrometry, 13, 1001-1008.

B. Kozlov, A. Saint, A. Skroce, 2003, Elemental fractionation in the formation of particulates, as observed by simultaneous isotopes measurement using laser ablation ICP-oa-TOFMS, Journal of Analytical Atomic Spectrometry, 18, 1069-1075.

A. Leach, G.M. Hieftje, 2001, Standardless semiquantitative analysis of metals using single-shot laser ablation inductively coupled plasma time-of-flight mass spectrometry, Analytical Chemistry, 73, 2959-2967.

C. Liu, X.L. Mao, S.S. Mao, X. Zeng, R. Greif, R.E. Russo, 2004, Nanosecond and femtosecond laser ablation of brass: particulate and ICPMS measurements, Analytcial Chemistry, 76, 379-383.

M. Ponting, J.A. Evans, V. Pashley, 2003, Fingerprinting Roman Mints Using Laser-Ablation MC-ICP-MS Lead Isotope Analysis, Archaeometry, 45, 591-597.

M. Rehkämper, A.N. Halliday, 1988, Accuracy and long-term reproducibility of lead isotopic measurements by multiple collector inductively coupled plasma mass spectrometry using an external method of mass discrimination, International Journal of Mass Spectrometry, 181, 123-133.

M. Rehkämper, K. Mezger, 2000, Investigations of matrix effects for Pb isotope ratio measurements by multiple collector ICP-MS: verification and application of optimized analytical protocols, Journal of Analytical Atomic Spectrometry, 15, 1451-1460.

M.K. Reuer, E.A. Boyle, B.C. Grant, 2003, Lead isotope analysis of marine carbonates and seawater by multiple collector ICP-MS, Chemical Geology, 200, 137-153.

J.F. Santos Zalduegui, S. García de Mandinabeitia, J.I. Gil Ibarguchi, 2004, A lead isotope database: the Los Pedroches- Alcudia Area (Spain); Implications for archaeometallurgical connections across southwestern and southeastern Iberia, Archaeometry, 46, 625-634.

Z.A. Stos-Gale, N.H. Gale, J. Houghton, R. Speakman, 1995, Lead isotope analysis of ores from the Western Mediterranean, Archaeometry, 37, 407-415.

M.S. Walton, 2004, A Materials Chemistry Investigation of Archaeological Lead Glazes, Unpublished D.Phil. Thesis, University of Oxford, Linacre College.

G. Wannemacker, F. Vanhaecke, L. Moens, A. Van Mele, 2000, Lead isotopic and elemental analysis of copper alloy statuettes by double focusing sector field ICP mass spectrometry, Journal of Analytical Atomic Spectrometry, 15, 323-327.

S. Willie, Z. Mester, R. Sturgeon, 2005, Isotope ratio precision with transient sample introduction using ICP orthogonal acceleration time-of-flight mass spectrometry, Journal of Analytical Atomic Spectrometry, 20, 1358-1364.

J. Woodhead, 2002, A simple method for obtaining highly accurate Pb isotope data by MC-ICP-MS, Journal of Analytical Atomic Spectrometry, 17, 1381-1385.

L. Yang, S. Willie, R. Sturgeon, 2005, Ultra-trace determination of mercury in water by cold-vapor generation isotope dilution mass spectrometry, Journal of Analytical Atomic Spectrometry, 20, 1226-1231.

W. Zhu, E.W.B. De Leer, M. Kennedy, P. Kelderman, G.J.F. Alaerts, 1997, Study of Partial Least-squares regression model for rare earth elemental determination of inductively coupled plasma mass spectrometry, Journal of Analytical Atomic Spectrometry, 12, 661-665.

List of Authors

Patrick Degryse, Katholieke Universiteit Leuven,
Patrick.Degryse@ees.kuleuven.be

Julian Henderson, Nottingham University,
Julian.Henderson@nottingham.ac.uk

Gregory Hodgins, University of Arizona,
ghodgins@physics.arizona.edu

Ian C. Freestone, Cardiff University,
Freestonei@cardiff.ac.uk

Sophie Wolf

Matthew Thirlwall, Royal Holloway College, University of London,
m.thirlwall@gl.rhul.ac.uk

Jane Evans, NIGL, British Geological Survey.

Youssef Barkoudah, Syrian European University.

Jens C. Schneider, Katholieke Universiteit Leuven,
Jens.Schneider@ees.kuleuven.be

Veerle Lauwers, Katholieke Universiteit Leuven,
Veerle.Lauwers@arts.kuleuven.be

Bernard Van Daele,
ligustinus@hotmail.com

Marleen Martens, VIOE,
marleen.martens@rwo.vlaanderen.be

Hans (D.J.) Huisman, RACM,
H.Huisman@racm.nl

David De Muynck, Katholieke Universiteit Leuven,
David.Demuynck@ees.kuleuven.be

Philippe Muchez, Katholieke Universiteit Leuven,
Philippe.Muchez@ees.kuleuven.be

Andrew J. Shortland, Cranfield University,
ashortland@cranfield.ac.uk

David Dungworth, English Heritage,
David.Dungworth@english-heritage.org.uk

Paz Marzo, University of Zaragoza.

Jozefina Pérez-Arantegui, University of Zaragoza,
jparante@posta.unizar.es

Francisco Laborda, University of Zaragoza.

Marc S. Walton, Getty Conservation Institute,
mwalton@getty.edu

The Editors

Patrick Degryse is Research Professor of Archaeometry at the department of Earth and Environmental Sciences and the Centre for Archaeological Sciences of the Katholieke Universiteit Leuven (Belgium). His research focuses on the ancient use of mineral resources and on the technology of craft production.

Julian Henderson is Professor of Archaeological Science at the Department of Archaeology, School of Humanities, at the University of Nottingham, UK. His research has broadly been concerned with the archaeological and scientific characteristics of ancient materials, especially vitreous materials and ceramics, and focuses on ancient technology in ancient social, economic and political contexts.

Greg Hodgins is an Assistant Research Scientist and an Assistant Professor of Anthropology at the National Science Foundation - Arizona Accelerator Mass Spectrometry Laboratory, University of Arizona, Tucson, USA. He has worked at the AMS laboratory at Oxford University and at a new AMS laboratory at the University of Georgia. His interests include compound-specific isotope analysis, environmental isotope analysis and archaeology.